T0189406

Lecture Notes in Computer Science 14533

Founding Editors

Gerhard Goos
Juris Hartmanis

Editorial Board Members

The series Lecture Notes in Computer Science (LNCS), including its subseries Lecture Notes in Artificial Intelligence (LNAI) and Lecture Notes in Bioinformatics (LNBI), has established itself as a medium for the publication of new developments in computer science and information technology research, teaching, and education.

LNCS enjoys close cooperation with the computer science R & D community, the series counts many renowned academics among its volume editors and paper authors, and collaborates with prestigious societies. Its mission is to serve this international community by providing an invaluable service, mainly focused on the publication of conference and workshop proceedings and postproceedings. LNCS commenced publication in 1973.

Anirban Mukhopadhyay · Ilkay Oksuz ·
Sandy Engelhardt · Dajiang Zhu · Yixuan Yuan
Editors

Deep Generative Models

Third MICCAI Workshop, DGM4MICCAI 2023
Held in Conjunction with MICCAI 2023
Vancouver, BC, Canada, October 8, 2023
Proceedings

Editors
Anirban Mukhopadhyay 🆔
TU Darmstadt
Darmstadt, Germany

Ilkay Oksuz 🆔
Istanbul Technical University
Istanbul, Türkiye

Sandy Engelhardt 🆔
University Hospital Heidelberg
Heidelberg, Germany

Dajiang Zhu
The University of Texas at Arlington
Arlington, TX, USA

Yixuan Yuan 🆔
City University of Hong Kong
Hong Kong, Hong Kong

ISSN 0302-9743   ISSN 1611-3349 (electronic)
Lecture Notes in Computer Science
ISBN 978-3-031-53766-0   ISBN 978-3-031-53767-7 (eBook)
https://doi.org/10.1007/978-3-031-53767-7

This Springer imprint is published by the registered company Springer Nature Switzerland AG
The registered company address is: Gewerbestrasse 11, 6330 Cham, Switzerland

Paper in this product is recyclable.

Preface

It was our genuine honor and great pleasure to organize the Third Workshop on Deep Generative Models for Medical Image Computing and Computer Assisted Intervention (DGM4MICCAI 2023), a satellite event at the 26th International Conference on Medical Image Computing and Computer Assisted Intervention (MICCAI 2023).

DGM4MICCAI was a single-track, half-day workshop consisting of high-quality, previously unpublished papers, presented orally (in a hybrid format), intended to act as a forum for computer scientists, engineers, clinicians and industrial practitioners to present their recent algorithmic developments, new results and promising future directions in Deep Generative Models. Deep generative models such as Diffusion Models, Generative Adversarial Networks (GAN) and Variational Auto-Encoders (VAE) are currently receiving widespread attention from not only the computer vision and machine learning communities, but also in the MIC and CAI community. These models combine advanced deep neural networks with classical density estimation (either explicit or implicit) to achieve state-of-the-art results. As such, DGM4MICCAI provided an all-round experience for deep discussion, idea exchange, practical understanding and community building around this popular research direction.

This year's DGM4MICCAI was held on October 8, 2023, in Vancouver, Canada. There was an enthusiastic response to the call for papers this year. We received 43 submissions for the workshop. Each paper was reviewed by at least three reviewers and we ended up with 23 accepted papers (53% acceptance rate) for the workshop. The accepted papers present fresh ideas on broad topics ranging from methodology (causal inference, latent interpretation, generative factor analysis etc.) to applications (mammography, vessel imaging, surgical videos etc.).

The high quality of the scientific program of DGM4MICCAI 2023 was due first to the authors who submitted excellent contributions and second to the dedicated collaboration of the international Program Committee and the other researchers who reviewed the papers. We would like to thank all the authors for submitting their valuable contributions and for sharing their recent research activities.

We are particularly indebted to the Program Committee members and to all the external reviewers for their precious evaluations, which permitted us to organize this event. We were also very pleased to benefit from the keynote lecture of the invited speaker, Ke Li, Simon Fraser University, Vancouver, Canada. We would like to express our sincere gratitude to these renowned experts for making the second workshop a successful platform to advance Deep Generative Models research within the MICCAI context.

August 2023   Anirban Mukhopadhyay
                                                                    Ilkay Oksuz
                                                                    Sandy Engelhardt
                                                                    Dajiang Zhu
                                                                    Yixuan Yuan

Organization

Organizing Committee

Anirban Mukhopadhyay	Technische Universität Darmstadt, Germany
Ilkay Oksuz	Istanbul Technical University, Turkey
Dajiang Zhu	University of Texas at Arlington, USA
Yixuan Yuan	City University of Hong Kong, China
Sandy Engelhardt	University Hospital Heidelberg, Germany

Program Committee

Li Wang	University of Texas at Arlington, USA
Tong Zhang	Peng Cheng Laboratory, China
Ping Lu	Oxford University, UK
Roxane Licandro	Medical University of Vienna, Austria
Veronika Zimmer	TU München, Germany
Dwarikanath Mahapatra	Inception Institute of AI, UAE
Michael Sdika	CREATIS, France
Jelmer Wolterink	University of Twente, The Netherlands
Alejandro Granados	King's College London, UK
Jinglei Lv	University of Sydney, Australia
Onat Dalmaz	Bilkent University, Turkey
Camila González	Stanford University, USA
Magda Paschali	Stanford University, USA

Student Organizers

Lalith Sharan	University Hospital Heidelberg, Germany
Henry Krumb	Technische Universität Darmstadt, Germany
Moritz Fuchs	Technische Universität Darmstadt, Germany
John Kalkhof	Technische Universität Darmstadt, Germany
Yannik Frisch	Technische Universität Darmstadt, Germany
Amin Ranem	Technische Universität Darmstadt, Germany
Caner Özer	Istanbul Technical University, Turkey

Additional Reviewers

Martin Menten
Soumick Chatterjee
Arijit Patra
Giacomo Tarroni
Angshuman Paul
Meilu Zhu
Antoine Sanner
Mariano Cabezas
Yipeng Hu
Salman Ul Hassan Dar
Shuo Wang
Christian Desrosiers
Chenyu Wang
Prateek Prasanna
Lorenzo Tronchin
Li Wang

Alberto Gomez
Mohamed Akrout
Shuo Wang
Despoina Ioannidou
Pedro Sanchez
Ninon Burgos
Haoteng Tang
Dorit Merhof
Xinyu Liu
Wei Peng
Qishi Yang
Matthias Wödlinger
Luis Garcia-Peraza Herrera
Meng Zhou
Yipeng Hu
Meilu Zhu

Contents

Methods

Privacy Distillation: Reducing Re-identification Risk of Diffusion Models

Virginia Fernandez[1]([✉]) [ID], Pedro Sanchez[2] [ID], Walter Hugo Lopez Pinaya[1] [ID], Grzegorz Jacenków[2] [ID], Sotirios A. Tsaftaris[2] [ID], and M. Jorge Cardoso[1] [ID]

[1] King's College London, London WC2R 2LS, UK
virginia.fernandez@kcl.ac.uk
[2] The University of Edinburgh, Edinburgh EH9 3FG, UK

Abstract. Knowledge distillation in neural networks refers to compressing a large model or dataset into a smaller version of itself. We introduce *Privacy Distillation*, a framework that allows a generative model to teach another model without exposing it to identifiable data. Here, we are interested in the privacy issue faced by a data provider who wishes to share their data via a generative model. A question that immediately arises is *"How can a data provider ensure that the generative model is not leaking patient identity?"*. Our solution consists of (i) training a first diffusion model on real data; (ii) generating a synthetic dataset using this model and filter it to exclude images with a re-identifiability risk; (iii) training a second diffusion model on the filtered synthetic data only. We showcase that datasets sampled from models trained with Privacy Distillation can effectively reduce re-identification risk whilst maintaining downstream performance.

Keywords: Privacy · Diffusion Models · Distillation

1 Introduction

Synthetic data have emerged as a promising solution for sharing sensitive medical image data [15]. By utilizing generative models, artificial data can be created with statistical characteristics similar to the original training data, thereby overcoming privacy, ethical, and legal issues that data providers face when sharing healthcare data [15,20,32]. Recent advancements have made generative models achieve sufficient quality to accurately represent the original medical data in terms of both realism and diversity [4]. Diffusion probabilistic models [4,10] (DPMs), are a new class of deep generative models which has been successfully applied to medical image data [17], and can scale particularly well to high resolution and 3D images [24]. However, the high performance of these models also raises a growing concern on whether they preserve privacy [5].

Supplementary Information The online version contains supplementary material available at https://doi.org/10.1007/978-3-031-53767-7_1.

Patient privacy is indeed a crucial concern for these models, as the aim is to share the synthetic datasets and the models themselves across institutions or make them open source [5]. Sharing trained generative models, in particular, can be useful[1] for fine-tuning with smaller datasets, performing anomaly detection [23,28], or even using synthetic data for segmentation [8]. Deep generative models are, however, prone to leak information about their training datasets [13]. A major risk in medical imaging is the potential for patient re-identification from the training dataset [16,32], especially when sharing models derived from private or protected datasets. *Re-identification risk* refers to the probability that a generative model synthesises an image which can be traced back to a real patient [22]. While the notion of what constitutes a person's identity may be ambiguous, it is possible to train deep learning models to accurately determine whether images belong to the same patient [22]. DPMs have shown to be particularly susceptible to attacks extracting its training data [3,30], more than other architectures such as generative adversarial networks (GANs) [3], but few publications have tackled solutions for DPM privacy preservation.

A common solution which allows privately sharing deep learning models with guarantees is differential privacy (DP) [1,7,16]. Generation of images with DP, however, has shown poor scalability to high resolution images. Existing DP methods [7] typically generate images at 32×32 resolution whereas latent diffusion models (LDM) can scale up to 1024×1024 [27] or 3D [24] images. At the same time, many discriminative tasks such detecting pathologies from X-ray require images at 224×224 resolution [6]. Recent work has shown that duplications and outliers datapoints are prone to be memorised [3], suggesting that filtering the training dataset might improve privacy. In [21], Packhäuser et al. filtered a dataset generated using a DPM according to a re-identification metric in order to reduce its re-identification risk. These synthetic datasets can still perform relatively well in downstream tasks [21]. Therefore, while reducing the re-identification risk for synthetic datasets is possible, sharing *private models* capable of generating high-resolution images tailored to the user's requirements remains an open challenge.

In this work, we propose a method for sharing models with a reduced re-identification risk via *privacy distillation*. In the distillation procedure, two diffusion models are trained sequentially. The first model is trained on real data and used to generate a synthetic dataset. Subsequently, the synthetic dataset is filtered by a re-identification network to eliminate images that could potentially be used to re-identify real patients while mainting usefulness in a downstream task. A second model is then trained on the filtered synthetic dataset, thus avoiding the risk of memorisation of the real images and subsequent potential re-identification of patients. The efficacy of the distilled model is evaluated by assessing the performance of a downstream classifier on the synthetic data generated by the distilled model.

---

[1] As seen in Stable Diffusion's successful public release followed by over 6 million downloads (by March 2023) of its weights by the community https://stability.ai/blog/stable-diffusion-public-release.

Contributions. (i) We train a conditional LDM on text-image pairs from a Chest X-ray dataset [4]; (ii) We assess re-identification risk of LDMs trained with different dataset sizes as well as how risk varies when the model is trained from scratch as opposed to fine-tuned; (iii) We verify that the distilled model has lower re-identification risk, whilst retaining information about the original dataset useful for classifiers on its generated data.

2   Methods

2.1   Data

We evaluate our method on a real dataset $D_{real} = \{(\mathbf{x}_i, \mathbf{c}_i) \mid \forall i \in (1, 2, \ldots, N)\}$ of images $\mathbf{x}$ and text conditions $\mathbf{c}$. We use images and radiological reports from the MIMIC-CXR 2.0.0 database [14]. As text, we use each report's "impression" section, which corresponds to an interpretative summary of the findings for supporting medical decisions. Following RoentGen [4], we filter the data used in this study based on the length of impression in tokens, which should not exceed 76 tokens due to the text encoder limit. Ultimately we obtained a set of 45,453 images belonging to 25,852 patients, each associated with an impression of the original radiological report. We split these into a train set of 23,268 patients (40,960 images) and a test set of 2,132 patients (3,101 images). 10% of the patients left for testing had half of their image and report pairs moved to the training dataset to allow us to assess re-identification when the patient, but not the query image, is part of the training dataset.

2.2   Diffusion Models

Diffusion models [10] learn to reverse a sequential image noising process, thus learning to map a pure noise image into a target data distribution. Diffusion models ϵ_θ can, therefore, be used as a generative model. We follow RoentGen [4] in training/fine-tuning a latent diffusion model (LDM) [27] pre-trained[2] on a subset of the LAION-5B database [29]. The LDM allows the generation of high-dimensional high-resolution images by having a diffusion model over the latent space of a variational autoencoder. The latent model is conditioned using a latent space derived from CLIP text encoder [25]. Here, we only fine-tune/train from scratch the diffusion model weights, leaving the autoencoder and CLIP text encoder as pre-trained [4]. We generate images using classifier-free guidance [11] with the PNDM scheduler [19]. Whenever we mention samples from an unconditional model, we refer to images generated with prompts from empty strings.

---

[2] https://huggingface.co/runwayml/stable-diffusion-v1-5.

2.3   Privacy Distillation

As we are interested in safely sharing the weights of generative models in a privacy-preserving manner, a major concern is that synthetic images generated by a model can be used to re-identify a patient from the real training dataset. Therefore, we propose an algorithm for training a diffusion model over filtered synthetic images, minimising the model's exposure to re-identifiable data.

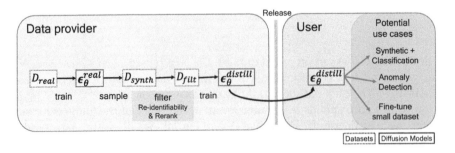

Fig. 1. Privacy Distillation Pipeline.

The procedure for *Privacy Distillation*, as depicted in Fig. 1, consists of the following steps: (i) Train a diffusion model $\epsilon_\theta{}^{real}$ on real data D_{real}; (ii) Generate a synthetic data D_{synth}; (iii) Filter D_{synth}, ensuring that none of the images are re-identifiable, to obtain $D_{filtered}$; (iv) Train a diffusion model $\epsilon_\theta{}^{distill}$ on $D_{filtered}$; (v) Share $\epsilon_\theta{}^{distill}$.

2.4   Filtering for Privacy

Identity Memorisation.   We hypothesise that synthetic images can enable re-identification due to model memorisation. A $\mathbf{x}_i$ is considered **memorised** by ϵ_θ if $\ell(\hat{\mathbf{x}}_i, \mathbf{x}_i) \geq \delta$ [3,32], where ℓ is a similarity function, δ is a threshold, and $\mathcal{A}$ is an algorithm which can extract an image $\hat{\mathbf{x}}_i$ from a generative model ϵ_θ without access to the original $\mathbf{x}_i$, $\hat{\mathbf{x}}_i = \mathcal{A}(\epsilon_\theta)$. We consider $\mathcal{A}$ to be a sampling algorithm with access to $\mathbf{c}_i$.

Defining an appropriate $\ell(\hat{\mathbf{x}}, \mathbf{x})$ allows controlling which aspects of the original data one wishes to measure for memorisation. Previous work [3] searches near-identical images utilising a Euclidean distance or pixel-by-pixel correspondence. Measuring *identity*, however, can be challenging and specific to certain modalities or organs [18], limiting the validity of such approaches.

Assessing Re-identification.   Instead of pixel-based [3] or structural-based [18] similarities, we measure identity with a deep model $\ell = f_\theta^{re-id}$, introduced by Packhäuser et al. [22]. The model is trained to classify images as belonging to the same patient or not. This model, devised for X-Ray images, consists of a siamese neural network with a ResNet-50 backbone. The model takes in two images, fuses the representation of the two branches and outputs a *re-identification score* (after a sigmoid) that we will note as s_{re-id}. If $s_{re-id} \geq \delta$ for a pair of synthetic and real images $(\hat{\mathbf{x}}, \mathbf{x})$, $\hat{\mathbf{x}}$ has been re-identified by ϵ_θ. For a set of synthetic images, we call *re-identification ratio* R_{re-id} the number of re-identified samples divided by the number of total samples. We explored the effect of varying δ and found no relevant difference in the resulting score for thresholds between 0.05 and 0.90, so we picked an intermediate threshold of $\delta = 0.5$.

We train f_θ^{re-id} from scratch on our training set, sampling positive (images from the same patient) and negative pairs. To avoid data imbalance, positive pairs, which were on average ten times less frequent, were oversampled, resulting in an effective dataset size of 472,992. We tested it on a set of even 101,592 non-repeated pairs, achieving a 99.16% accuracy (AUC 0.9994).

Retrieval. For data sampled without conditioning, we also utilised a retrieval model $f_\theta^{retrieval}$ proposed in [22]. The model is a siamese neural network with an architecture similar to f_θ^{re-id}. However, the $f_\theta^{retrieval}$ excludes the layers from the merging point onwards, to function solely as a feature extractor. To identify the closest image in terms of identity, we computed the Euclidean distance between the embeddings of the query sampled image and every image in our training set. When evaluating pairs of the test set from the real dataset, our trained model obtained high mean average precision at R (mAP@R) of about 95% and a high Precision@1 (the precision when evaluating how many times the top-1 images in the retrieved lists are relevant) of 97%. This way, approach enabled us to analyze and evaluate the unconditioned synthetic data accurately.

Constrative Reranking. We ensure that the images in $D_{filtered}$ used for training $\epsilon_\theta^{distill}$ correspond to their conditioning. Therefore, we rerank the synthetic images in $D_{filtered}$ based on the image alignment with the conditioning, similar to Dall-E [26]. We leverage a contrastive text-image model f_θ^{im2tex} pre-trained on MIMIC-CXR [2]. An alignment score $s_{align} = f_\theta^{im2tex}(\hat{\mathbf{x}}_i, \mathbf{c}_i)$ is computed between an image and a text prompt by passing them through an image and text encoder respectively and taking the cosine similarity between their latent spaces.

Filtering Strategy. We generate N_c synthetic images for each prompt $\mathbf{c}_i$ in D_{real}. We choose $N_c = 10$; although a higher value would have been desirable, increasing it bore a certain computational cost. Therefore, D_{synth} has $N_c * N$ elements. We compute s_{align} between all generated images and corresponding conditioning using f_θ^{im2tex}; and s_{re-id} between the generated images and the real image corresponding to its prompt. For unconditional models, we use $f_\theta^{retrieval}$ to find the strongest candidate in the dataset before computing s_{re-id}. We remove all re-identified images $s_{re-id} \geq \delta$ and choose, for each $\mathbf{c}_i$, the synthetic image with the highest s_{align}.

2.5   Downstream Task

To assess the quality of synthetic datasets, we train a classifier f_θ^{class} of 5 different pathologies (Cardiomegaly, Edema, Consolidation, Atelectasis, Pleural Effusion) based on the model ranked first in the CheXpert Stanford ML leaderboard [33].[3] We trained a DenseNet121 on our datasets and tested it in the real hold-out test set. The network is pre-trained for 5 epochs on cross-entropy loss, then trained for another 5 epochs on an AUC loss, as per [33] (Table 1).

Table 1. Evaluating the influence of pre-training and conditioning on ϵ_θ^{real}.

Pre-trained	Conditional	$s_{re-id} \downarrow$	FID $\downarrow$
–	–	0.057 ± 0.232	54.56
–	✓	0.015 ± 0.124	81.95
✓	-	0.034 ± 0.181	97.91
✓	✓	0.022 ± 0.232	79.27

3   Experiments

First, we evaluate how/when identity memorisation happens and the effect of training the model and sampling under different conditions and dataset sizes. Then we showcase that our model trained under Privacy Distillation can be used to train a downstream classification model.

3.1   Measuring Re-identification Risk of Latent Diffusion Models

Effect of Fine-Tuning. We explored the differences in terms of s_{align} between fine-tuning the model pre-trained on LAION-5B or training from scratch, and between sampling using conditioning or not. For the conditioned generation, we sample 100 instances for each of the first 400 prompts of the training dataset, resulting in 40,000 samples. For the unconditional generation, we sample 40,000 images and use the retrieval network to get the closest images in the training dataset. We calculate the re-identification ratio and evaluate the quality of the images using the Fréchet inception distance (FID) score [9], based on the features extracted by the pre-trained DenseNet121 from the *torchxrayvision* package [6]. The lowest re-identification ratio was achieved for the data sampled from a model trained from scratch using conditioning. Unconditionally-generated datasets have higher re-identification ratios but achieve a better FID score. Nonetheless, their usability is limited, as conditional sampling allows the user to guide the generation process.

Effect of Training Dataset Size on Memorisation. We trained one model on our full training dataset, and three models on 1%, 10% and 50% of the training dataset, respectively. Then, we calculated the R_{re-id} of a set of samples inferred using 100 instances of 400 training prompts (40,000 images in total). Figure 2

---

[3] We used their code, available at https://github.com/Optimization-AI/LibAUC.

shows the *re-identification ratio*, defined as the average number of times a sample was re-identified divided by the total of generated samples. Supplementary Fig. 1 shows examples of re-identified and non re-identified samples.

As opposed to findings in the literature, where bigger training set sizes result in less leakage [3], the re-identification ratio was lowest for the model trained on only 1% of the data. The TSNE plots of Fig. 2 suggest that re-identification tends to happen in specific clusters [31]. We looked at the radiological reports of the top 10 most re-identified prompts, and 90% of them were associated with similar pathological phenotypes: atelectasis, pleural effusion, and lung opacity. In parallel, we observed that the proportion of images associated with these pathologies is less frequent in the first 1% of the data than it is in the whole dataset, which suggests that generated data from models trained on the full dataset might be more re-identifiable due to overfitting to subject-specific pathological features.

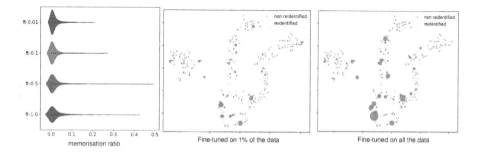

Fig. 2. Left: violin plots showing the distribution of the average re-identification ratio for the models fine-tuned in different portions of the training dataset; the middle and right plots are TSNE plots of the f_θ^{im2tex} embeddings of the first 400 prompts used to test this experiment, for the model trained on 1% and 100% of the data, respectively. Orange dots correspond to prompts for which none of the generated 100 instances was re-identified, whereas blue dots are associated with prompts re-identified to some extent, the size being proportional to the re-identification ratio. (Color figure online)

Effect of Filtering. We compare the impact of our filtering strategy on the synthetic datasets D_{synth} and $D_{filtered}$. We sample 10 instances for each of the 40,959 training prompts from the proposed privacy-preserving model. We then filter by s_{re-id}, and pick the sample with the better s_{align} score, resulting in a filtered dataset of 40,959 images (Fig. 3). We found that filtering improves the s_{align} and reducing the number of memorised (re-identifiable) samples to 0, given a δ.

3.2 Privacy Distillation

We now show empirically that $\epsilon_\theta^{distill}$ indeed reduces re-identification risk. We also assess whether $\epsilon_\theta^{distill}$ is able to produce useful synthetic datasets for downstream tasks. We train classifiers f_θ^{class} (see Sect. 2.5) on D_{real}, D_{synth}, capped

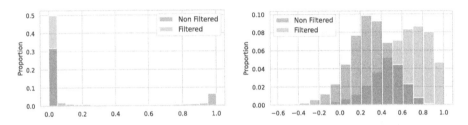

Fig. 3. *Left*: s_{re-id} distribution; *right*: s_{align} distribution. Comparing distribution of the non-filtered D_{synth} and filtered $D_{filtered}$ dataset.

at 40,000 images, and a dataset $D_{distill}$ of 40,000 images sampled from $\epsilon_\theta{}^{distill}$, our distilled model, using the training set prompts. We measure the AUC on the 3,101 images of our test set. The results are reported in Table 2, in addition to the re-identification ratio R_{re-id} and the s_{align} score. Supplementary Fig. 2 shows example images sampled from $\epsilon_\theta{}^{distill}$.

Training a model on synthetic data slightly affects performance (AUC and s_{align} decrease), but the resulting value is still comparable to the literature [12]. Nonetheless, the re-identification ratio between the initial and the distilled models is decreased by more than 3-fold. We hypothesise that filtering re-identifiable data might also filter out unique phenotypes, more prone to be memorised [3], resulting in reduced model generalisability and performance. Another possible reason is a loss of quality of the images produced by the model after the distillation process. On the other hand, the fact that $D_{distill}$ is generated by applying a threshold on R_{re-id}, along with the inherent model stochasticity, could explain why R_{re-id} doesn't reach 0%.

Table 2. Privacy Distillation performance: memorisation ratio, predicted AUC for the classifier on the real test set and s_{align} score.

Dataset	R_{re-id} ↓	f_θ^{class} AUC ↑	s_{align} ↑
D_{real}	–	0.863	$0.698_{0.259}$
D_{synth}	4.24%	0.830	$0.645_{0.219}$
$D_{distill}$	1.34%	0.810	$0.611_{0.272}$

4   Discussion and Conclusion

This study has demonstrated that the application of *privacy distillation* can effectively reduce the risk of re-identification in latent diffusion models without excessively compromising the downstream task performance. Our main goal is to enable sharing of private medical imaging data via generative models. In line with other approaches [7], there is a trade-off between privacy (degree of filtering) and utility (downstream performance). However, differently from DP techniques [7], which can only be performed over low-resolution datasets, we train our private model on dataset at 256×256 resolution.

Although in this work we evaluated the impact of important factors such as dataset size and fine-tuning, our work is still limited to an imaging type and a re-identification metric. Nonetheless, the principle of *privacy distillation* could potentially be extended to other modalities, conditioning types or re-identification metrics. Further work should also assess the effect of applying it iteratively (i.e. more filtering-sampling-training steps) and use other downstream tasks (e.g. segmentation) for evaluation, to support the usability of the method.

References

1. Abadi, M., et al.: Deep learning with differential privacy. In: ACM SIGSAC, pp. 308–318 (2016)
2. Boecking, B., et al.: Making the most of text semantics to improve biomedical vision-language processing. In: Avidan, S., Brostow, G., Cissé, M., Farinella, G.M., Hassner, T. (eds.) ECCV 2022. LNCS, vol. 13696, pp. 1–21. Springer, Cham (2022). https://doi.org/10.1007/978-3-031-20059-5_1
3. Carlini, N., et al.: Extracting training data from diffusion models. arXiv (2023)
4. Chambon, P., et al.: RoentGen: vision-language foundation model for chest X-ray generation. arXiv preprint arXiv:2211.12737 (2022)
5. Chen, R.J., Lu, M.Y., Chen, T.Y., Williamson, D.F., Mahmood, F.: Synthetic data in machine learning for medicine and healthcare. Nat. Biomed. Eng. **5**, 6 (2021)
6. Cohen, J.P., et al.: TorchXRayVision: a library of chest X-ray datasets and models. In: MIDL (2022)
7. Dockhorn, T., Cao, T., Vahdat, A., Kreis, K.: Differentially private diffusion models (2022)
8. Fernandez, V., et al.: Can segmentation models be trained with fully synthetically generated data? In: Zhao, C., Svoboda, D., Wolterink, J.M., Escobar, M. (eds.) SASHIMI 2022. LNCS, vol. 13570, pp. 79–90. Springer, Cham (2022). https://doi.org/10.1007/978-3-031-16980-9_8
9. Heusel, M., Ramsauer, H., Unterthiner, T., Nessler, B., Hochreiter, S.: GANs trained by a two time-scale update rule converge to a local nash equilibrium. In: NeurIPs, vol. 30 (2017)
10. Ho, J., Jain, A., Abbeel, P.: Denoising diffusion probabilistic models. In: NeurIPS (2020)
11. Ho, J., Salimans, T.: Classifier-free diffusion guidance. In: NeurIPS 2021 Workshop
12. Jacenkow, G., O'Neil, A.Q., Tsaftaris, S.A.: Indication as prior knowledge for multimodal disease classification in chest radiographs with transformers. In: IEEE ISBI (2022)
13. Jegorova, M., et al.: Survey: leakage and privacy at inference time. IEEE Trans. Pattern Anal. Mach. Intell. **45**, 1–20 (2023)
14. Johnson, A.E., et al.: MIMIC-CXR, a de-identified publicly available database of chest radiographs with free-text reports. Sci. Data **6**, 317 (2019)

15. Jordon, J., Wilson, A., van der Schaar, M.: Synthetic data: Opening the data flood-gates to enable faster, more directed development of machine learning methods. arXiv preprint arXiv:2012.04580 (2020)
16. Kaissis, G.A., Makowski, M.R., Rückert, D., Braren, R.F.: Secure, privacy-preserving and federated machine learning in medical imaging. Nat. Mach. Intell. **2**, 305–311 (2020)
17. Kazerouni, A., et al.: Diffusion models for medical image analysis: a comprehensive survey. arXiv:2211.07804 (2022)
18. Kumar, K., Desrosiers, C., Siddiqi, K., Colliot, O., Toews, M.: Fiberprint: a subject fingerprint based on sparse code pooling for white matter fiber analysis. Neuroimage **158**, 242–259 (2017)
19. Liu, L., Ren, Y., Lin, Z., Zhao, Z.: Pseudo numerical methods for diffusion models on manifolds. In: ICLR (2022)
20. Murtaza, H., Ahmed, M., Khan, N.F., Murtaza, G., Zafar, S., Bano, A.: Synthetic data generation: state of the art in health care domain. Comput. Sci. Rev. **48**, 100546 (2023)
21. Packhäuser, K., Folle, L., Thamm, F., Maier, A.: Generation of Anonymous Chest Radiographs Using Latent Diffusion Models for Training Thoracic Abnormality Classification Systems (2022)
22. Packhäuser, K., Gündel, S., Münster, N., Syben, C., Christlein, V., Maier, A.: Deep learning-based patient re-identification is able to exploit the biometric nature of medical chest X-ray data. Sci. Rep. **12**(1), 1–13 (2022)
23. Pinaya, W.H.L., et al.: Fast unsupervised brain anomaly detection and segmentation with diffusion models. In: Wang, L., Dou, Q., Fletcher, P.T., Speidel, S., Li, S. (eds.) MICCAI 2022. LNCS, vol. 13438, pp. 705–714. Springer, Cham (2022). https://doi.org/10.1007/978-3-031-16452-1_67
24. Pinaya, W.H., et al.: Brain imaging generation with latent diffusion models. In: Mukhopadhyay, A., Oksuz, I., Engelhardt, S., Zhu, D., Yuan, Y. (eds.) DGM4MICCAI 2022. LNCS, vol. 13609, pp. 117–126. Springer, Cham (2022). https://doi.org/10.1007/978-3-031-18576-2_12
25. Radford, A., et al.: Learning transferable visual models from natural language supervision. In: ICML (2021)
26. Ramesh, A., et al.: Zero-shot text-to-image generation. In: ICML (2021)
27. Rombach, R., Blattmann, A., Lorenz, D., Esser, P., Ommer, B.: High-resolution image synthesis with latent diffusion models. In: Proceedings of the IEEE/CVF Conference on Computer Vision and Pattern Recognition, pp. 10684–10695 (2022)
28. Sanchez, P., Kascenas, A., Liu, X., O'Neil, A.Q., Tsaftaris, S.A.: What is healthy? Generative counterfactual diffusion for lesion localization. In: Mukhopadhyay, A., Oksuz, I., Engelhardt, S., Zhu, D., Yuan, Y. (eds.) DGM4MICCAI 2022. LNCS, vol. 13609, pp. 34–44. Springer, Cham (2022). https://doi.org/10.1007/978-3-031-18576-2_4
29. Schuhmann, C., et al.: LAION-5b: an open large-scale dataset for training next generation image-text models. In: NeurIPS Datasets and Benchmarks Track (2022)
30. Somepalli, G., Singla, V., Goldblum, M., Geiping, Wu, J., Goldstein, T.: Diffusion art or digital forgery? Investigating data replication in diffusion models. In: CVPR (2023)
31. Su, R., Liu, X., Tsaftaris, S.A.: Why patient data cannot be easily forgotten? In: Wang, L., Dou, Q., Fletcher, P.T., Speidel, S., Li, S. (eds.) MICCAI 2022. LNCS, vol. 13438, pp. 632–641. Springer, Cham (2022). https://doi.org/10.1007/978-3-031-16452-1_60

32. Yoon, J., Drumright, L.N., van der Schaar, M.: Anonymization through data synthesis using generative adversarial networks (ADS-GAN). IEEE J. Biomed. Health Inform. **24** (2020)
33. Yuan, Z., Yan, Y., Sonka, M., Yang, T.: Large-scale Robust Deep AUC Maximization: A New Surrogate Loss and Empirical Studies on Medical Image Classification

Federated Multimodal and Multiresolution Graph Integration for Connectional Brain Template Learning

Jia Ji and Islem Rekik$^{(\boxtimes)}$ (iD)

BASIRA Lab, Imperial-X and Department of Computing, Imperial College London,
London, UK
i.rekik@imperial.ac.uk
https://basira-lab.com/

Abstract. The connectional brain template (CBT) is an integrated graph that
normalizes brain connectomes across individuals in a given population. A *well-
centered* and *representative* CBT can offer a holistic understanding of the brain
roadmap landscape. Catchy but rigorous graph neural network (GNN) architectures were tailored for CBT integration, however, ensuring the privacy in CBT
learning from large-scale connectomic populations poses a significant challenge.
Although prior work explored the use of federated learning in CBT integration,
it fails to handle brain graphs at multiple resolutions. To address this, we propose a novel federated multi-modal multi-resolution graph integration framework
(Fed2M), where each hospital is trained on a graph dataset from modality m and
at resolution r_m to generate a local CBT. By leveraging federated aggregation
in a shared layer-wise manner across different hospital-specific GNNs, we can
debias the CBT learning process towards its local dataset and force the CBT to
move towards a global center derived from multiple private graph datasets *without compromising privacy*. Remarkably, the hospital-specific CBTs generated by
Fed2M converge towards a shared global CBT, generated by aggregating learned
mappings across heterogeneous federated integration GNNs (i.e., each hospital
has access to a specific unimodal graph data at a specific resolution). To ensure
the global centeredness of each hospital-specific CBT, we introduce a novel loss
function that enables global centeredness across hospitals and enforces consistency among the generated CBTs. Our code is available at https://github.com/
basiralab/Fed2M.

Keywords: Federating varying architectures · Multi-modal multi-resolution
population graph fusion · Connectional Brain Learning

1 Introduction

The connectome, which characterizes the interactions between brain regions of interest
(ROIs) at various scales [1], plays a crucial role in advancing network neuroscience

Supplementary Information The online version contains supplementary material available at
https://doi.org/10.1007/978-3-031-53767-7_2.

A. Mukhopadhyay et al. (Eds.): DGM4MICCAI 2023, LNCS 14533, pp. 14–24, 2024.
https://doi.org/10.1007/978-3-031-53767-7_2

[2]. To comprehensively map and examine these interactions, the development of a *well-centered* and *representative* connectional brain template (CBT) has gained attention in recent years [3,4]. The CBT serves as a tool to normalize brain networks of a population and capture the connected core or backbone within a complex manifold of graphs [3]. Moreover, it facilitates the comparison of core networks across different populations, enabling the differentiation between healthy and disordered populations [5,6]. By identifying deviations from the normalized brain network representation, it becomes possible to more effectively discern pathological changes occurring in brain networks [7]. This capability holds great potential for enhancing our understanding of various neurological conditions and disorders.

The high-dimensionality of brain connectivity data from different views, such as principle curvature, sulcal depth, and cortical thickness, adds complexity to constructing a CBT [8–10]. The primer Deep Graph Normalizer (DGN) [8] presented the first GNN model that aims to normalize and integrate a population of *multi-view* brain networks into a single CBT. It introduced the Subject Normalization Loss (SNL) to generate representative CBTs by leveraging complementary information from multiple view-based networks (MVBNs). However, DGN overlooks the importance of preserving topological integrity in the generated CBT. To address this, the multi-view graph normalizer network (MGN) proposes a topology loss that penalizes deviations from ground-truth brain networks with different views. Both DGN and MGN assume that graphs are derived from the same modality and at the same resolution. It is argued that constructing a multi-modal CBT by integrating structural, morphological, and functional brain networks offers a holistic perspective that enhances our understanding of brain connectivity [2,11,12]. To bridge this gap, the *Multi-modal Multi-resolution* Brain Graph Integrator Network (M2GraphIntegrator) was proposed [13]. M2GraphIntegrator employs resolution-specific autoencoders to map brain networks from various modalities and resolutions into a shared embedding space. These learned embeddings are then processed through a specialized integrator network to generate the CBT, while leveraging the heterogeneity of brain connectivity data (i.e., graphs) from multiple modalities and resolutions.

The strict cyberinfrastructure regulations within healthcare organizations, coupled with the risks of privacy breaches, pose significant challenges to the sharing of medical data [14,15]. Existing brain network integration methodologies typically focus on graph integration from a single global source [8,13,16–19]. The emerging field of federated learning offers a potential solution by enabling health data analytics while preserving privacy, particularly for hospitals that prioritize privacy protection and face limitations in the generalizability of results [20,21]. Fed-CBT approach [22] has emerged as a promising method for federated multi-view graph fusion into a CBT. Hospitals serve as clients, retaining their local training multi-view connective data and updating a shared model through aggregating locally-computed updates. The server performs layer-wise averaging using temporary-weighted averaging [23], and the resulting global weights are shared with hospitals for subsequent updates. Such iterative process enables collaborative learning while preserving data in a federated setting. Fed-CBT, although effective, is limited to evaluating *uni-modal* data and cannot integrate brain networks with multiple resolutions. To address this limitation, we propose *the first federated multi-modal multi-resolution graph integration* framework called Fed2M. In Fed2M, we set out to learn from heterogeneous hospitals, where each hospital has a brain graph data derived from one modality m (e.g., functional or morphological) and at a specific resolution r_m. Note that the local GNN models may vary across hospitals in architecture depending on the specific modality and resolution of their graph dataset. *First*, the local brain graph data from each hospital is passed through its corresponding autoencoder, mapping graphs from different modalities and at various resolutions into a shared embedding space. *Secondly,* batches of embeddings are processed through a batch-based normalizer to estimate batch-based CBTs. *Thirdly,* the weights of all local models are sent to a server and aggregated in a shared layer-wise manner. *Lastly,* for each hospital, the batch-based CBTs are debiased using element-wise median operation to generate the hospital-specific CBT. Notably, **we note that all hospital-specific CBTs are drawn towards a global center shared among all hospitals in a fully data-preserving and learnable way** (Suppl. Fig. 1). This allows for universal CBT learning from a locally limited data, where the universality is imposed by the federation process. Moreover, we introduce a novel loss that boosts the **global centeredness** of hospital-specific CBTs, (i.e., local CBTs). Fed2M offers a comprehensive solution for integrating *multi-modal* and *multi-resolution* brain networks while utilizing the benefits of federated learning.

2   Proposed Method

In this section, we provide a detailed description of the steps involved in our Fed2M[1] framework. Figure 1 illustrates the three key steps of the proposed method: A) Batch-based embeddings and batch-based CBTs generation. B) Shared layer-wise aggregation and batch-based embeddings sampling. C) Post-training: hospital-specific CBT generation.

A- Batch-Based Embeddings and Batch-Based CBTs Generation. In Fig. 1-A, each hospital is trained locally by first vectorizing its local brain graph data to form a local *graph population*. This graph population is then passed through a *self-mapper or autoencoder* to map it into a shared embedding space across hospitals. Subsequently, batches of embeddings are normalized using a *batch-based normalizer* to generate batch-based embedding from which the CBT matrix is derived and batch-based CBTs.

Graph Population. In our proposed federated pipeline with k hospitals denoted as $\{H_m\}_{m=1}^{k}$, each hospital has a population of brain networks derived from a specific modality m (e.g., functional or morphological) and at a particular resolution r_m. To facilitate batch-based normalization in the framework, all brain networks are processed in batches. Thus, we represent the connectivity data for each hospital as follows:

$$\mathcal{X}_m = \{X_{m,b}\}_{b=1}^{n_b}, X_{m,b} = \{X_{m,b}^s\}_{s=1}^{n_s^b},$$

where $X_{m,b}^s \in \mathbb{R}^{r_m \times r_m}$. Each $X_{m,b}^s \in \mathbb{R}^{r_m \times r_m}$ denotes a connectivity matrix (or adjacency matrix) of resolution r_m (i.e., number of ROIs) for subject s in batch b, and there

---

[1] https://github.com/basiralab/Fed2M

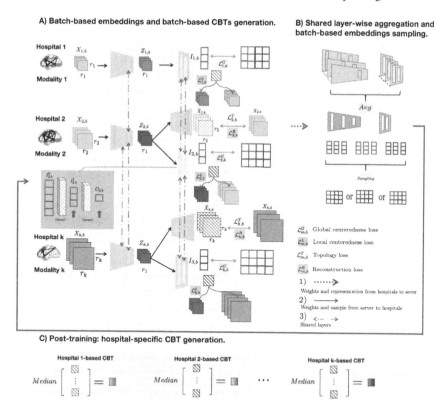

Fig. 1. *Proposed federated multi-modal multi-resolution graph integration framework (Fed2M) for learning universal connectional brain network.* **A)** *Batch-based embeddings and batch-based CBTs generation.* **B)** *Shared layer-wise aggregation and batch-based embeddings sampling.* **C)** *Post-training: hospital-specific CBT generation.*

are n_s^b subjects in batch b. $X_{m,b}$ represents the set of modality-specific brain networks in a batch b derived from modality m. $\mathcal{X}_m$ encompasses all the batches of modality m-specific brain networks. We note that $r_m \in \{r_1, r_2, \ldots, r_k\}$ which stipulates a learning paradigm from heterogeneous GNN architectures. To map the local brain connectivity dataset $\mathcal{X}_m$ into a lower-dimensional embedding space at the lowest resolution r_1, we utilize graph convolution (GCNConv), which encodes the node feature vectors and learns hidden layer representations [24]. Prior to applying GCNConv, we vectorize the connectivity matrices $X_{m,b}^s$. Each brain network is represented as a subject in a local population graph, where the connectivity matrix serves as the feature vector of the subject node denoted as $V_{m,b}^s \in \mathbb{R}^{1 \times \hat{r}_m}$, with $\hat{r}_m = \frac{r_m \times (r_m - 1)}{2}$.

Self-mapper and Hospital-Specific Autoencoder. To generate the embeddings $Z_{m,b}^s$ for each hospital's local feature vectors $V_{m,b}^s$, we initialize the hospital's GNN model using the self-mapper or resolution-specific graph autoencoders from the M2GraphIntegrator framework [13] (Fig. 1-A). Hospital 1, with the lowest resolution r_1 for its local brain data, utilizes the self-mapper E_1 to embed the feature vectors at the minimal resolution into the same resolution level. For hospitals with higher resolutions, resolution-specific or hospital-specific autoencoders E_m are used for initialization. Both architectures share a common structure, consisting of GCNblocks. Each GCNblock comprises a GCNlayer (or GCNConv) followed by a sigmoid function and a dropout layer. The self-mapper E_1 of H_1 is composed of a single GCNblock denoted as $\{l_1\}$. On the other hand, the hospital-specific encoders $\{E_m\}_{m=2}^{k}$ consist of three GCNblocks, namely $\{l_3^m, l_2, l_1\}$ from left to right (Fig. 1-A). In these encoders, l_1 is shared across all hospitals, while l_2 is shared among the hospitals, excluding the one with the minimal resolution ($\{H_m\}_{m=2}^{k}$). By passing the vectorized local connectivity matrices ($V_{m,b}^s$) through the corresponding self-mapper or encoder, the local feature vectors $V_{m,b}^s$ at the resolution $\hat{r}_m$ are mapped into a shared embedding space of size $\hat{r}_1$. This mapping process allows for the embeddings of hospital-specific brain networks at the same resolution level, facilitating further integration. We note that we federate the **shared** encoding and normalizing layers across varying GNNs.

Hospital-Specific Decoder. The hospital-specific encoder E_m is paired with its corresponding decoder D_m to reconstruct the ground-truth brain network using the embedding $Z_{m,b}^s$. However, the self-mapper is not paired with a decoder to preserve the structure of the minimal-resolution brain network while mapping it to the same resolution. Although the decoder is not directly involved in CBT integration, it plays a key role in ensuring that the embeddings retain essential traits of the ground-truth brain networks. The decoder D_m follows the same architecture as E_m, but in a reversed version. The reconstructed feature vector is denoted as $\widetilde{V}_{m,b}^s$, which can be antivectorized to matricial form $\widetilde{X}_{m,b}^s$. To ensure that the embedding $Z_{m,b}^s$ captures the essential traits of the ground-truth connectivity matrix, we incorporate the reconstruction loss and topology loss from the M2GraphIntegrator. The reconstruction loss $\mathcal{L}_{m,b}^R$, calculated in batches, measures the mean Frobenius distance (MFD) between the reconstructed connectivity matrix $\widetilde{X}_{m,b}^s$ and the ground-truth matrix $X_{m,b}^s$. Additionally, we employ the topology loss to ensure the topology soundness of the decoded connectivity matrix from the learnt embedding. The topology loss $\mathcal{L}_{m,b}^T$, also calculated in batches, measures the absolute difference between the node strength vectors [25] of the reconstructed connectivity matrix $P_{m,b}^s$ and the ground-truth matrix $\widetilde{P}_{m,b}^s$.

Batch-Based Normalizer. To normalize the embeddings within a batch $Z_{m,b}$, we introduce a batch-based normalizer N_m to generate batch-based CBTS, which is shared across all hospitals. The normalizer comprises two blocks, denoted as $\{l_4, l_5\}$, arranged from left to right (Fig. 1-A). Each block consists of a linear layer followed by a sigmoid function. First, the embeddings $Z_{m,b} \in \mathbb{R}^{n_s^b \times \hat{r}_m}$ are linearized into a linear vector $I_{m,b}^0 \in \mathbb{R}^{1 \times i_{m,b}}$, where $i_{m,b} = n_s^b \times \hat{r}_m$. Then, $I_{m,b}^0$ is passed through the batch-based normalizer, and the resulting vector is antivectorized to produce a batch-based CBT denoted as $C_{m,b} \in \mathbb{R}^{r_1 \times r_1}$. The embedding obtained from l_4 is represented as $I_{m,b}^1$ (batch-based embedding). We introduce a *local centeredness loss* $\mathcal{L}_{m,b}^L$ to ensure that the generated batch-based CBT captures not only the traits of brain networks within the current batch but also represents the entire population of hospital-specific local brain network data. This is achieved by comparing the batch-based CBT $C_{m,b}$ of batch b against randomly selected N_L embedded connectivity matrices from other batches through clustering techniques such as hierarchical clustering. This evaluates the centeredness of the CBT against other sample batches. In addition to identifying a well-centered CBT within each hospital's population, which we refer to as "local centeredness", our ultimate objective is to achieve the convergence of hospital-specific CBTs towards an implicit global CBT (global centeredness). To achieve this, we introduce a novel *global centeredness loss* $\mathcal{L}_{m,b}^G$. This loss is computed as the Euclidean distance between $I_{m,b}^1$ and randomly selected N_G batch-based embeddings $I_{m',b'}^1$ from other hospitals. In our federated learning framework, hospitals exchange batch-based embeddings with each other through the server, facilitating collaborative learning. The shared embeddings offer benefits while maintaining privacy, as each client has no knowledge of the architectures used by other clients. Furthermore, the absence of shared decoders under federation reduces the risk of reconstructing local brain networks of other clients. This approach ensures data privacy and security while promoting the integration of heterogeneous and multi-resolution brain connectivity data. In our federated learning framework, hospitals exchange batch-based embeddings with each other through the server. The collaborative exchange enables hospitals to benefit from the shared batch-based embeddings without directly accessing the sensitive information within their local data. Additionally, by processing embeddings in batches and utilizing the first layer of batch normalizer, it reduces the risk of reconstructing personal brain networks.

The training process is conducted in rounds, with each round comprising a predefined number of training epochs. Within each round, the trained hospitals undergo the backward propagation phase, followed by the transmission of the updated model weights and batch-based embeddings to the server at the end. In summary, we define the total batch-specific loss for a training batch b as follows:

$$
\mathcal{L}_{m,b} = \underbrace{\sum_{s=1}^{n_s^b} \left\| X_{m,b}^s - \tilde{X}_{m,b}^s \right\|_2^2}_{\mathcal{L}_{m,b}^R} + \lambda_1 \underbrace{\sum_{s=1}^{n_s^b} \left\| P_{m,b}^s - \tilde{P}_{m,b}^s \right\|_1}_{\mathcal{L}_{m,b}^T}
$$

$$
+ \lambda_2 \underbrace{\left(\sum_{s=1}^{n_s^b} \sum_{s'=1}^{N_L} \left\| C_{m,b} - Z_{m,b}^s \right\|_2^2 + \left\| C_{m,b} - Z_{m,b'}^{s'} \right\|_2^2 \right)}_{\mathcal{L}_{m,b}^L} + \lambda_3 \underbrace{\sum_{b'=1}^{N_G} \left\| I_{m,b} - I_{m',b'} \right\|_2}_{\mathcal{L}_{m,b}^G}
$$

B- Shared Layer-Wise Aggregation and Batch-Based Embeddings Sampling.
Figure 1-B illustrates the operations performed at the server, which involve aggregating the local model weights and sampling the batch-based embeddings received from the local hospitals. Once the server receives the models' weights and batch-based embeddings from the hospitals, it performs FedAvg [23] on the shared layers' weights. Specifically, the weights of l_1, l_2, l_4, l_5 are averaged as follows:

$$
W_{l_1}^{t+1} = \frac{1}{k} \sum_{m=1}^{k} W_{m,l_1}^t, \, W_{l_i}^{t+1} = \frac{1}{(k-1)} \sum_{m=2}^{k} W_{m,l_i}^t, \, \text{where } i \in \{2,4,5\}
$$

Here, W_{m,l_1}^t and W_{m,l_i}^t represents the weights of shared layers of hospital m during the tth round. The federated weights for the subsequent round, denoted as $W_{l_1}^{t+1}$ and

$W_{l_i}^{t+1}$, are then sent back to the local models of the hospitals. As a result, the weights of the local models are updated as follows:

$$W_{m,l_1}^{t+1} = W_{l_1}^{t+1}, \text{where } m \in \{1,\dots,k\}$$

$$W_{m,l_i}^{t+1} = W_{l_i}^{t+1}, \text{where } m \in \{2,\dots,k\}, i \in \{2,4,5\}$$

Furthermore, the server randomly selects batch-based embeddings from other hospitals m' and sends them to hospital m. This allows hospital m to update its global centeredness loss for the $(t+1)^{\text{th}}$ round by comparing its batch-based embeddings with the sampled batch-based embeddings received from the server at the end of the tth round.

C- Post-training: Hospital-Specific CBT Generation. To mitigate the influence of specific batches and promotes a more representative and robust CBT for each hospital, we propose a post-training step involving an element-wise median operation on the batch-based CBTs (Fig. 1-C). Specifically, we calculate the median of the batch-based CBTs as $C_m = median\,[C_{m,1}, C_{m,2}, \dots, C_{m,n_b}]$, where n_b is the total number of batches in the local population. C_m is the estimated local CBT of hospital m.

3   Results and Discussion

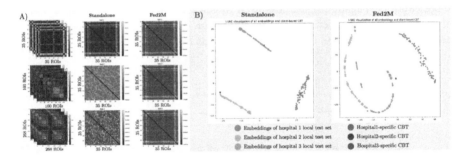

Fig. 2. *The learnt hospital-specific CBTs and t-SNE plot based on different strategies.* A) Learnt hospital-specific CBTs by standalone and Fed2M models. B) t-SNE plot of brain graph embeddings and hospital-specific CBTs by different strategies.

Dataset. We conducted training and evaluation using the Southwest University Longitudinal Imaging Multimodal (SLIM) Brain Data Repository [26]. The SLIM dataset comprises 279 healthy subjects, each depicted by 3 connectomes at the following resolutions: 35, 160, and 268. Connectomes represented with 35 ROIs are based on morphological networks extracted from T1-weighted MRIs, whereas the ones with 160 and 268 ROIs are derived from resting-state functional MRI (rs-fMRI) and correspond to functional networks. To ensure generalizability, we perform a 4-fold cross-validation to split the dataset into training and testing sets. To simulate the federated learning framework, we created three virtual hospitals, each with access to different types of MRIs.

Hospital 1 had access to morphological MRIs with 35 ROIs, hospital 2 had access to functional MRIs with 160 ROIs, and hospital 3 had access to functional MRIs with 268 ROIs.

Benchmarks and Hyperparameter Tuning. We evaluated our model using three variants for comparison: a **standalone** model without federation, an **ablated** version of Fed2M without global centeredness loss optimization, and **Fed2M**. To optimize the local centeredness loss of the hospital-specific local CBT, we employed two different sampling algorithms for randomly selecting the local training subsets in the SNL loss. Furthermore, we trained each variant of the model with two different numbers of updates per round: 1 epoch per round and 10 epochs per round. This allowed us to assess the performance and effectiveness of our proposed Fed2M framework under various configurations and settings. We initialized two clusters for hierarchical sampling and selected 5 training subjects at each epoch. The hyperparameters λ_1, λ_2, and λ_3 in our loss function $\mathcal{L}_{m,b}$ were set to 2, 0.5, and 0.2, respectively, using a grid search strategy. We used a learning rate of 0.00005 for all 3 hospitals and the Adam optimizer.

Table 1. *Evaluation of Fed2M method using various measures.* Sampling method for the calculation of local centeredness loss: hierarchical clustering (H) and random (R) sampling. EPR represents epochs per round. Consistency evaluates the consistency of the self-mapper. Reconstruction MAE and Topology end evaluates the reconstruction capability and topology soundness of autoencoder. Local centeredness and global centeredness of hospital-specific CBTs are evaluated as well. The bold number represents the best performing method, while the underlined number indicates the second best. Fed2M significantly outperforms ablated Fed2M (*p-value* < 0.05). Both Fed2M and ablated Fed2M outperform the standalone approach in terms of the local centeredness of H 1 and H 2 (*p-value* < 0.001). H 1: hospital 1. H 2: hospital 2. H 3: hospital 3.

	EPR	Model variation	Consistency	Local Centeredness			Reconstruction MAE		Topology Snd.		Global Centeredness
				H 1	H 2	H 3	H 2	H 3	H 2	H 3	
H	10	Standalone	**0.84312**	0.03050	**0.00739**	0.04009	0.15153	0.18109	0.20414	**0.19300**	0.28080
		Ablated Fed2M	0.84478	0.01670	0.07633	**0.01672**	0.15141	0.18266	0.20418	0.19312	0.00366
		Fed2M	0.84476	**0.01662**	0.02453	0.01680	0.15150	**0.18225**	**0.15212**	0.19313	**0.00202**
	1	Standalone	**0.84313**	0.03042	**0.00738**	0.04029	0.15153	0.18175	0.20414	0.19303	0.27982
		Ablated Fed2M	0.84460	0.02174	0.02826	**0.02168**	0.15140	0.17928	0.20418	0.19302	0.00677
		Fed2M	0.84458	**0.02206**	0.02886	0.02151	0.15149	0.17967	**0.20407**	0.19312	**0.00379**
R	10	Standalone	**0.84330**	0.02919	**0.00740**	0.03444	0.15153	0.18178	0.20414	**0.19310**	0.26270
		Ablated Fed2M	0.84478	0.01624	0.02318	**0.01587**	0.15150	0.18224	**0.20407**	0.19313	0.00360
		Fed2M	0.84480	**0.01617**	0.02337	0.01620	0.15140	0.18289	0.20418	0.19314	**0.00191**
	1	Standalone	**0.84330**	0.02919	**0.00740**	0.03444	0.15153	0.18178	0.20414	0.19310	0.26270
		Ablated Fed2M	0.84462	**0.02136**	0.02738	0.02210	0.15140	0.17949	0.20418	0.19307	0.00685
		Fed2M	0.84461	0.02162	0.02741	**0.02126**	0.15149	0.17959	**0.20407**	0.19312	**0.00317**

Evaluation. The reconstruction capability of the hospital-specific autoencoder is assessed by calculating the mean absolute difference between the ground-truth and reconstructed connectivity matrices. Additionally, the preservation of topological properties is evaluated by the difference between the node strength of the ground-truth and reconstructed connectivity matrices. We also measure the consistency of the self-mapper by comparing the input minimal-resolution brain network with its corresponding output embedding. To assess the *local centeredness* of the hospital-specific CBT, we compute the MFD between the hospital-CBT generated from the local training set and the embeddings of test set. To ensure global centeredness, we calculate the Euclidean distance between hospital-specific CBTs. Hospitals 1 and 3 exhibit lower local centeredness in both ablated Fed2M and Fed2M (p-$value$ < 0.001), while hospital 2 shows higher local centeredness (Table 1). This disparity arises from the bias in CBT learning towards the local dataset [27, 28]. Since our ultimate goal is to achieve *global centeredness*, we anticipate that the hospital-specific CBTs may not be locally well-centered but contribute to a more comprehensive, representative, and globally centered brain network. Remarkably, Fed2M generates more globally centered hospital-specific CBTs compared to ablated Fed2M (p-$value$ < 0.05). Figure 2-A displays estimated hospital-specific CBTs, while Fig. 2-B shows the blending of functional MRI embeddings and distinct morphological MRI embeddings, demonstrating Fed2M's capability to preserve modality-specific traits and achieve convergence towards a global center.

4   Conclusion

In this paper, we presented a primer federated multi-modal multi-resolution integration framework from heterogeneous multi-source brain graphs. Our framework includes the introduction of the global centeredness loss to optimize the hospital-specific CBTs, ensuring the multi-modal information capability. By leveraging federated learning, we demonstrated that hospital-specific CBT derived by our approach captures shared traits among populations from different modalities and resolutions, leading to a more comprehensive understanding of brain connectivity while maintaining data privacy. However, a key limitation of this work is the potential privacy leakage of batch-based embeddings. In our future work, we plan to address this limitation by incorporating differential privacy techniques.

References

1. Sporns, O., Tononi, G., Kötter, R.: The human connectome: a structural description of the human brain. PLoS Comput. Biol. **1**, e42 (2005)
2. Bassett, D., Sporns, O.: Network neuroscience. Nat. Neurosci. **20**, 353–364 (2017)
3. Rekik, I., Li, G., Lin, W., Shen, D.: Estimation of brain network atlases using diffusive-shrinking graphs: application to developing brains. In: Niethammer, M., et al. (eds.) IPMI 2017. LNCS, vol. 10265, pp. 385–397. Springer, Cham (2017). https://doi.org/10.1007/978-3-319-59050-9_31
4. Chaari, N., Akdağ, H., Rekik, I.: Comparative survey of multigraph integration methods for holistic brain connectivity mapping. Med. Image Anal. **85**, 102741 (2023)
5. Wassermann, D., Mazauric, D., Gallardo-Diez, G., Deriche, R.: Extracting the core structural connectivity network: guaranteeing network connectedness through a graph-theoretical approach. In: Ourselin, S., Joskowicz, L., Sabuncu, M.R., Unal, G., Wells, W. (eds.) MICCAI 2016, Part I. LNCS, vol. 9900, pp. 89–96. Springer, Cham (2016). https://doi.org/10.1007/978-3-319-46720-7_11
6. Uylings, H., Rajkowska, G., Sanz-Arigita, E., Amunts, K., Zilles, K.: Consequences of large interindividual variability for human brain atlases: converging macroscopical imaging and microscopical neuroanatomy. Anat. Embryol. **210**, 423–31 (2006)

7. Heuvel, M., Sporns, O.: A cross-disorder connectome landscape of brain dysconnectivity. Nat. Rev. Neurosci. **20**, 1 (2019)

8. Gurbuz, M.B., Rekik, I.: Deep graph normalizer: a geometric deep learning approach for estimating connectional brain templates. In: Martel, A.L., et al. (eds.) MICCAI 2020, Part VII. LNCS, vol. 12267, pp. 155–165. Springer, Cham (2020). https://doi.org/10.1007/978-3-030-59728-3_16

9. Van Essen, D., Glasser, M.: The human connectome project: progress and prospects. In: Cerebrum: the Dana Forum on Brain Science, vol. 2016 (2016)

10. Verma, J., Gupta, S., Mukherjee, D., Chakraborty, T.: Heterogeneous edge embedding for friend recommendation. In: Azzopardi, L., Stein, B., Fuhr, N., Mayr, P., Hauff, C., Hiemstra, D. (eds.) ECIR 2019, Part II. LNCS, vol. 11438, pp. 172–179. Springer, Cham (2019). https://doi.org/10.1007/978-3-030-15719-7_22

11. Acosta-Mendoza, N., Gago-Alonso, A., Carrasco-Ochoa, J., Martínez-Trinidad, J.F., Pagola, J.: Extension of canonical adjacency matrices for frequent approximate subgraph mining on multi-graph collections. Int. J. Pattern Recognit Artif Intell. **31**, 1750025 (2017)

12. Bunke, H., Riesen, K.: Recent advances in graph-based pattern recognition with applications in document analysis. Pattern Recogn. **44**, 1057–1067 (2011)

13. Cinar, E., Haseki, S.E., Bessadok, A., Rekik, I.: Deep cross-modality and resolution graph integration for universal brain connectivity mapping and augmentation. In: Manfredi, L., et al. (eds.) ISGIE GRAIL 2022. LNCS, vol. 13754, pp. 89–98. Springer, Cham (2022). https://doi.org/10.1007/978-3-031-21083-9_9

14. Jin, H., Luo, Y., Li, P., Mathew, J.: A review of secure and privacy-preserving medical data sharing. IEEE Access **7**, 61656–61669 (2019)

15. Malin, B., Emam, K., O'Keefe, C.: Biomedical data privacy: problems, perspectives, and recent advances. J. Am. Med. Inform. Assoc. JAMIA **20**, 2–6 (2012)

16. Wang, B., et al.: Similarity network fusion for aggregating data types on a genomic scale. Nat. Methods **11**, 333–337 (2014)

17. Mhiri, I., Rekik, I.: Joint functional brain network atlas estimation and feature selection for neurological disorder diagnosis with application to autism. Med. Image Anal. **60**, 101596 (2019)

18. Mhiri, I., Mahjoub, M.A., Rekik, I.: Supervised multi-topology network cross-diffusion for population-driven brain network atlas estimation. In: Martel, A.L., et al. (eds.) MICCAI 2020, Part VII. LNCS, vol. 12267, pp. 166–176. Springer, Cham (2020). https://doi.org/10.1007/978-3-030-59728-3_17

19. Gürbüz, M., Rekik, I.: MGN-Net: a multi-view graph normalizer for integrating heterogeneous biological network populations. Med. Image Anal. **71**, 102059 (2021)

20. Xu, J., Glicksberg, B.S., Su, C., Walker, P., Bian, J., Wang, F.: Federated learning for healthcare informatics. J. Healthc. Inform. Res. **5**, 1–19 (2021)

21. Rieke, N., et al.: The future of digital health with federated learning. NPJ Digit. Med. **3**, 119 (2020)

22. Bayram, H.C., Rekik, I.: A federated multigraph integration approach for connectional brain template learning. In: Syeda-Mahmood, T., et al. (eds.) ML-CDS 2021. LNCS, vol. 13050, pp. 36–47. Springer, Cham (2021). https://doi.org/10.1007/978-3-030-89847-2_4

23. McMahan, B., Moore, E., Ramage, D., Hampson, S., Arcas, B.A.: Communication-efficient learning of deep networks from decentralized data. In: Artificial Intelligence and Statistics, PMLR, pp. 1273–1282 (2017)

24. Kipf, T.N., Welling, M.: Semi-supervised classification with graph convolutional networks. arXiv preprint arXiv:1609.02907 (2016)

25. Barrat, A., Barthelemy, M., Pastor-Satorras, R., Vespignani, A.: The architecture of complex weighted networks. Proc. Natl. Acad. Sci. U.S.A. **101**, 3747–52 (2004)

26. Wei, L., et al.: Longitudinal test-retest neuroimaging data from healthy young adults in Southwest China. Sci. Data **4**, 170017 (2017)
27. Hanzely, F., Richtárik, P.: Federated learning of a mixture of global and local models. arXiv preprint arXiv:2002.05516 (2020)
28. Ji, S., Pan, S., Cambria, E., Marttinen, P., Yu, P.: A survey on knowledge graphs: representation, acquisition, and applications. IEEE Trans. Neural Netw. Learn. Syst. **33**, 494–514 (2021)

Metrics to Quantify Global Consistency in Synthetic Medical Images

Daniel Scholz[1,2]([✉]), Benedikt Wiestler[2], Daniel Rueckert[1,3], and Martin J. Menten[1,3]

[1] Lab for AI in Medicine, Technical University of Munich, Munich, Germany
daniel.scholz@mri.tum.de
[2] Department of Neuroradiology, Klinikum rechts der Isar, Technical University of Munich, Munich, Germany
[3] BioMedIA, Department of Computing, Imperial College London, London, UK

Abstract. Image synthesis is increasingly being adopted in medical image processing, for example for data augmentation or inter-modality image translation. In these critical applications, the generated images must fulfill a high standard of biological correctness. A particular requirement for these images is global consistency, i.e an image being overall coherent and structured so that all parts of the image fit together in a realistic and meaningful way. Yet, established image quality metrics do not explicitly quantify this property of synthetic images. In this work, we introduce two metrics that can measure the global consistency of synthetic images on a per-image basis. To measure the global consistency, we presume that a realistic image exhibits consistent properties, e.g., a person's body fat in a whole-body MRI, throughout the depicted object or scene. Hence, we quantify global consistency by predicting and comparing explicit attributes of images on patches using supervised trained neural networks. Next, we adapt this strategy to an unlabeled setting by measuring the similarity of implicit image features predicted by a self-supervised trained network. Our results demonstrate that predicting explicit attributes of synthetic images on patches can distinguish globally consistent from inconsistent images. Implicit representations of images are less sensitive to assess global consistency but are still serviceable when labeled data is unavailable. Compared to established metrics, such as the FID, our method can explicitly measure global consistency on a per-image basis, enabling a dedicated analysis of the biological plausibility of single synthetic images.

Keywords: Generative Modeling · Synthetic Images · Image Quality Metrics · Global Consistency

1 Introduction

With recent improvements in deep learning-based generative modeling [4,9,25], image synthesis is increasingly utilized in medical image processing. It has seen

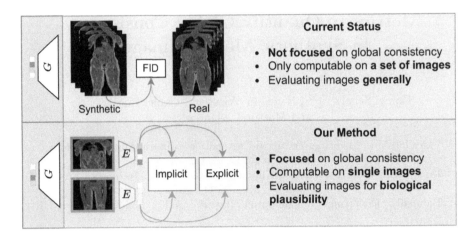

Fig. 1. We present a novel method to quantify global consistency in generated images. Most established image quality metrics, like FID [8], are not designed to measure the biological correctness of medical images. Conversely, our approach measures the global consistency of synthetic medical images, like whole-body MRIs, based on their explicit and implicit features.

application in inter-modality transfer [18], counterfactual image generation [20], anomaly detection [6], data augmentation [28], and synthetic dataset generation [22]. When using synthetic images in critical medical systems, it is vital to ensure the biological correctness of the images. One crucial aspect of image realism is its global consistency [5,12,31]. Global consistency refers to an image's overall coherence and structure so that all parts of the image fit together in a realistic and plausible way. While several others have researched methods to improve the global consistency of synthetic images [11,16], these works do not quantitatively assess the global consistency of these images in a standardized fashion. This is because existing metrics, such as Inception Score [24], Fréchet Inception Distance (FID) [8], and Precision and Recall [15,23], only measure image quality in terms of fidelity and variety.

In this work, we introduce solutions to measure the global consistency of synthetic images (Fig. 1). To this end, we make the following contributions.

- We propose an approach to quantify global consistency by determining attributes on different image regions. We call this method *explicit* quantification of global consistency.
- Next, we adapt this approach to a setting in which explicit labels are not available. To this end, we utilize the cosine similarity between feature vectors of patches in the image as a global consistency measure. These *implicit* features are predicted by neural networks trained in a self-supervised fashion.
- In extensive experiments, we compare our proposed metrics with FID, one of the most established image quality metrics, with regard to its ability to measure global consistency in synthetic images. We perform our experiments

on the challenging task of whole-body magnetic resonance image (MRI) synthesis, in which it is crucial that the various depicted body parts match.

2 Related Works

2.1 Global Consistency in Image Synthesis

The notion of global consistency in image synthesis has been researched in computer vision. Multiple important works [12,31] describe synthesizing complex images with multiple objects as challenging and lacking global coherence. Integrating the attention mechanism [30] into the GAN architecture [5,11] facilitates generating more globally consistent images. To evaluate their adherence to the properties in the real data, Hudson *et al.* [11] statistically compare property co-occurrences in the generated images, similar to [28]. The use of large autoregressive models advances the generation of ultra-high-resolution images while maintaining global consistency [16]. They use a block-wise FID to assess the quality of individual blocks in the image, which only evaluates the realism of individual patches but does not measure the global consistency within a single image. In summary, none of these works have dedicated quantitative metrics for global consistency.

2.2 Metrics Measuring Quality of Generated Images

Several metrics, such as Inception Score [24], Fréchet Inception Distance (FID) [8], and Precision and Recall [15,23], have been proposed in the literature to assess the quality of synthetic images. The most established metric, the FID [8], measures image quality and variation in a single value by comparing the distribution over features from sets of real and synthetic images. Multiple variants have been proposed in the literature to address the limitations of FID. These variants focus on overcoming the bias towards a large number of samples [1,3], the lack of spatial features [29] or standardization of its calculation [21]. However, the global consistency remains, at most, only validated as part of general image fidelity.

Zhang *et al.* [32] measure a learned perceptual image patch similarity (LPIPS) between patches of two separate images. While this metric is conceptually similar to ours, their work focuses on evaluating different kinds of representations and similarity measures between two images for their correspondence to human judgment. However, they do not assess global consistency within a single image. Sun *et al.* [28] evaluate their hierarchical amortized GAN by quantifying the accuracy of clinical predictions on synthetic images. Their evaluation strategy only compares statistics over the clinical predictions between real and synthetic data but does not incorporate per-image analysis. In general, existing metrics do not explicitly address the quantification of global consistency.

2.3 GANs for Whole-Body MRI Synthesis

Only few works have researched the challenging task of generating synthetic whole-body MRIs. Mensing *et al.* [19] adapt a FastGAN [17] and a StyleGAN2

[10,14] to generate whole-body MRIs. They primarily evaluate their generated images using the Fréchet Inception Distance (FID) [8]. However, they do not focus on assessing global consistency of the synthetic images.

3   Method

We propose two novel metrics to measure the global consistency of synthetic images. We distinguish between *implicit* and *explicit* quantification of global consistency, which are described in the following (see Fig. 2).

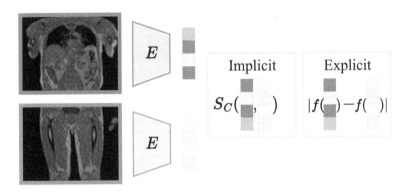

Fig. 2. Two strategies to assess the global consistency of an image based on the feature representations of the superior and inferior half of the body. *Explicit*: Absolute error between an explicit attribute predicted from the feature representation using some regression head f. *Implicit*: Cosine similarity S_C between the feature representations.

3.1   Explicit Quantification

Our method for explicitly quantifying global consistency is based on the notion that biological properties should be consistent in different parts of a person's body. For example, a person's body mass index (BMI) should be similar when viewing the superior part of a whole-body MRI depicting the torso and the inferior part containing the legs. To assess its global consistency, we compare various biological attributes, such as age, body fat percentage, or BMI, in two parts of the synthetic images. While individual organs might age at different rates [26], our method assumes that the overall age of the superior part and inferior part of a person's body still contain consistent age-related information. In addition, the body fat mass between the limbs and the trunk correlates and can hence serve as marker for consistency in a synthetic image [13]. We generate two views of the whole-body MRI by simply cropping the superior and inferior half of the image. Other possible cropping modes include random cropping, cropping based on semantic regions, such as organs, and grid-structured cropping. The two views are each evaluated using dedicated *referee neural networks*. We train several neural networks in a supervised fashion to predict one of three different biological properties for either the superior or inferior image view.

By comparing the predicted attributes via the absolute error, we can obtain a proxy measure for the global consistency of a synthetic image. For a more holistic analysis, we simultaneously compare an average error of all biological attributes.

3.2   Implicit Quantification

Detailed annotations of the data are not always available, rendering supervised training of referee networks infeasible. Therefore, we propose the use of implicit features extracted via a network that has been trained via self-supervision as an alternative measure for global consistency.

As before, we crop two views from the synthetic image and extract one feature vector for each view by applying an encoder network. Here, the encoder network is trained using SimCLR [2], a self-supervised contrastive learning framework alleviating the need for labels during training. SimCLR is trained to return similar representations for two views of the same image and diverging representations for two views of different images. The similarity between the embedding of the two views is obtained by calculating their cosine similarity. To calculate a global consistency measure for a given image, we obtain the cosine similarity between the embeddings of the superior and inferior views.

3.3   Experimental Setup

We conduct experiments using 44205 whole-body MRIs from the UK Biobank population study [27], which we split into 36013 training images, 4096 validation images, and 4096 test images. We extract the slice along the coronal plane in the intensity center of mass of the 3d volumes and normalize them to the range of $[0, 1]$. We train one ResNet50 [7] network per attribute on the training set as a referee network for the explicit quantification experiments. We also fit a ResNet50 using SimCLR [2] to our training images to extract features for the implicit quantification strategy.

The validation images are used to evaluate the accuracy of the referee networks for the explicit quantification strategy. We find that the networks achieve good performance on the attribute estimation. The mean absolute error (MAE) for age estimation is 3.9 years $\pm$ 2.98 years on the superior half and 4.4 years $\pm$ 3.35 years on the inferior half. Similarly, we achieve an MAE of 0.97 $\pm$ 0.83 on the superior and 1.11 $\pm$ 0.93 on the inferior half for BMI estimation and 2.10% $\pm$ 1.70% on the superior and 2.36% $\pm$ 1.89% on the inferior half for body fat percentage prediction. Ultimately, we compare the variation in biological properties of the explicit metric, the cosine similarity of the implicit metric, and the FID on all test set images.

4   Results

4.1   Distinguishing Consistent from Inconsistent Images

Initially, we analyze the two proposed metrics on a dataset of consistent and
inconsistent images. We construct the inconsistent images by stitching the superior part and inferior part of two different whole-body MRIs from the test set
together (see Fig. 3). The sharp edge at the seam of the inconsistent images is
a very distinctive feature. In order to avoid the metrics being influenced by it,
we remove the central 10% of both the consistent and inconsistent images. We
compare our two metrics with the FID [8], which is calculated using two distinct
sets of images. One half of the consistent images serves as the reference dataset
for calculating the FID of either the other half of the consistent images or the
inconsistent images, respectively.

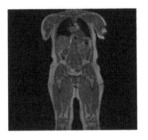 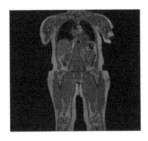 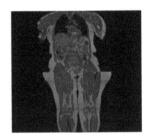

Fig. 3. A comparison of an original whole-body MRI (left) with the modified versions
used in our experiments, i.e., consistent (middle) and inconsistent (right) superior-
inferior combinations with the central 10% removed.

Our metrics differentiate well between consistent and inconsistent images (see
Table 1, top). For the explicit strategy, we report the mean over the superior-
inferior errors of age, BMI, and body fat percentage prediction after normalizing
them to a range between 0 and 1. While the FID is also influenced by global
consistency, our metric distinguishes more clearly between consistent and inconsistent.

We present a detailed analysis of the explicit attribute errors in Fig. 4. The
experiment shows that body fat percentage and BMI are more distinctive biological attributes than age.

Additionally, we investigate the correlation between our implicit and explicit
metrics to verify the utility of the implicit strategy in the absence of labels (see
Fig. 5). These findings suggest the potential utility of the implicit quantification
strategy as a weaker alternative to explicit quantification.

4.2   Global Consistency in Synthetic Whole-Body MRIs

We conduct an exemplary assessment of global consistency on 1000 synthetic
images using our implicit and explicit metrics and the FID. The synthetic whole-
body MRIs were generated using a StyleGAN2 [14] that we trained on images of

Table 1. Comparison of our explicit global consistency metrics, implicit global consistency metric, and FID in two different experiments. In the first one, we calculate all metrics on constructed consistent and inconsistent images. In the second experiment, the metrics are compared for real and synthetic datasets, akin to the envisioned use case of our proposed method.

Dataset	FID ($\downarrow$)	Explicit (Ours, $\downarrow$)	Implicit (Ours, $\uparrow$)
Consistent	**14.10**	**0.09 ± 0.05**	**0.59 ± 0.12**
Inconsistent	16.10	0.24 ± 0.11	0.37 ± 0.17
Real	**14.10**	0.09 ± 0.050	**0.59 ± 0.12**
Synthetic	17.13	0.09 ± 0.049	0.55 ± 0.14

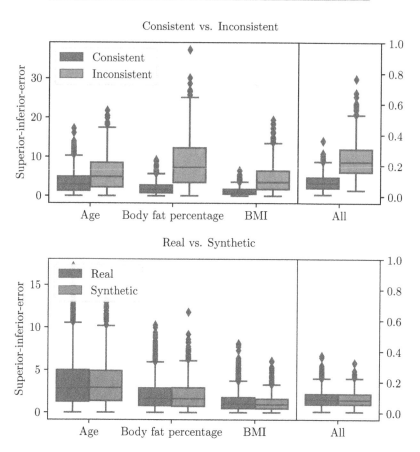

Fig. 4. The per-attribute results of the explicit absolute errors between the superior and inferior part of the consistent and inconsistent images (top) and real and synthetic images (bottom). The rightmost column: an average over the 0-1-normalized per-attribute errors.

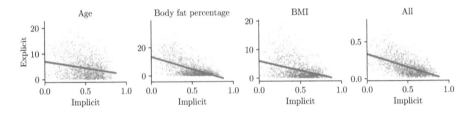

Fig. 5. The correlations between our implicit and explicit metrics verifying the utility of the implicit strategy in the absence of labels on real images.

the UK Biobank [27]. The results suggest an overall high global consistency and a low error in biological attributes in the synthetic images (Table 1, bottom). The images show overall high fidelity to the real images due to the comparable FID to the real images.

Our metrics differ only slightly between real and synthetic in the per-attribute analysis (see Fig. 4, bottom). The high values in our metrics indicate a high degree of global consistency in the synthetic images.

5   Discussion and Conclusion

In this work, we have proposed two strategies to quantify the global consistency in synthetic medical images. We found that global consistency influences established metrics for synthetic image quality, such as the FID, yet the differences between consistent and inconsistent images are more pronounced in our novel metrics. Our first metric explicitly quantifies the error between predicted biological attributes in the superior and inferior half of a single whole-body MR image. However, this approach relies on labels to train neural networks that determine the biological attributes. As a solution, we also presented a second metric based on implicit representations that can be obtained via a self-supervised trained network. Both strategies have proven suitable for assessing synthetic medical images in terms of their biological plausibility.

We envision that our work will complement the existing landscape of image quality metrics - especially in medical imaging - and that it will be used to develop and benchmark generative models that synthesize globally consistent medical images. An extension of our work to the 3D domain is theoretically simple but may be practically challenging due to the additional complexity when training SimCLR for the implicit and the referee networks for the explicit metric. Moreover, global consistency analysis for other image modalities and use cases can be enabled through retraining the feature extraction networks on domain specific data with corresponding augmentations. Ultimately, we believe our work can potentially increase the trust in using synthetic data for critical medical applications in the future.

Acknowledgments. This research has been conducted using the UK Biobank Resource under Application Number 87802. This work is funded by the Munich Center for Machine Learning. We thank Franz Rieger for his valuable feedback and discussions.

References

1. Bińkowski, M., Sutherland, D.J., Arbel, M., Gretton, A.: Demystifying MMD GANs. In: International Conference on Learning Representations (2018)
2. Chen, T., Kornblith, S., Norouzi, M., Hinton, G.: A simple framework for contrastive learning of visual representations. In: International Conference on Machine Learning, pp. 1597–1607. PMLR (2020)
3. Chong, M.J., Forsyth, D.: Effectively unbiased FID and inception score and where to find them. In: Proceedings of the IEEE Computer Society Conference on Computer Vision and Pattern Recognition, pp. 6069–6078. IEEE Computer Society (2020)
4. Dhariwal, P., Nichol, A.: Diffusion models beat GANs on image synthesis. Adv. Neural. Inf. Process. Syst. **34**, 8780–8794 (2021)
5. Esser, P., Rombach, R., Ommer, B.: Taming transformers for high-resolution image synthesis. In: Proceedings of the IEEE/CVF Conference on Computer Vision and Pattern Recognition, pp. 12873–12883 (2021)
6. Han, C., et al.: MADGAN: unsupervised medical anomaly detection GAN using multiple adjacent brain MRI slice reconstruction. BMC Bioinform. **22**(2), 1–20 (2021)
7. He, K., Zhang, X., Ren, S., Sun, J.: Deep residual learning for image recognition. In: Proceedings of the IEEE Conference on Computer Vision and Pattern Recognition, pp. 770–778 (2016)
8. Heusel, M., Ramsauer, H., Unterthiner, T., Nessler, B., Hochreiter, S.: GANs trained by a two time-scale update rule converge to a local nash equilibrium. In: Advances in Neural Information Processing Systems, vol. 30 (2017)
9. Ho, J., Jain, A., Abbeel, P.: Denoising diffusion probabilistic models. Adv. Neural. Inf. Process. Syst. **33**, 6840–6851 (2020)
10. Hong, S., et al.: 3D-StyleGAN: a style-based generative adversarial network for generative modeling of three-dimensional medical images. In: Engelhardt, S., et al. (eds.) DGM4MICCAI/DALI 2021. LNCS, vol. 13003, pp. 24–34. Springer, Cham (2021). https://doi.org/10.1007/978-3-030-88210-5_3
11. Hudson, D.A., Zitnick, C.L.: Generative adversarial transformers. In: Proceedings of the 38th International Conference on Machine Learning, ICML 2021 (2021)
12. Johnson, J., Gupta, A., Fei-Fei, L.: Image generation from scene graphs. In: Proceedings of the IEEE Conference on Computer Vision and Pattern Recognition, pp. 1219–1228 (2018)
13. Jung, S., Park, J., Seo, Y.G.: Relationship between arm-to-leg and limbs-to-trunk body composition ratio and cardiovascular disease risk factors. Sci. Rep. **11**(1), 17414 (2021)
14. Karras, T., Laine, S., Aittala, M., Hellsten, J., Lehtinen, J., Aila, T.: Analyzing and improving the image quality of styleGAN. In: Proceedings of the IEEE/CVF Conference on Computer Vision and Pattern Recognition, pp. 8110–8119 (2020)
15. Kynkäänniemi, T., Karras, T., Laine, S., Lehtinen, J., Aila, T.: Improved precision and recall metric for assessing generative models. In: Advances in Neural Information Processing Systems, vol. 32 (2019)

16. Liang, J., et al.: NUWA-infinity: autoregressive over autoregressive generation for infinite visual synthesis. In: Oh, A.H., Agarwal, A., Belgrave, D., Cho, K. (eds.) Advances in Neural Information Processing Systems (2022)

17. Liu, B., Zhu, Y., Song, K., Elgammal, A.: Towards faster and stabilized GAN training for high-fidelity few-shot image synthesis. In: International Conference on Learning Representations (2021)

18. Liu, Y., et al.: CT synthesis from MRI using multi-cycle GAN for head-and-neck radiation therapy. Comput. Med. Imaging Graph. **91**, 101953 (2021)

19. Mensing, D., Hirsch, J., Wenzel, M., Günther, M.: 3D (c) GAN for whole body MR synthesis. In: Mukhopadhyay, A., Oksuz, I., Engelhardt, S., Zhu, D., Yuan, Y. (eds.) DGM4MICCAI 2022. LNCS, vol. 13609, pp. 97–105. Springer, Cham (2022). https://doi.org/10.1007/978-3-031-18576-2_10

20. Menten, M.J., et al.: Exploring healthy retinal aging with deep learning. Ophthalmol. Sci. **3**(3), 100294 (2023)

21. Parmar, G., Zhang, R., Zhu, J.Y.: On aliased resizing and surprising subtleties in GAN evaluation. In: CVPR (2022)

22. Pinaya, W.H., et al.: Brain imaging generation with latent diffusion models. In: Mukhopadhyay, A., Oksuz, I., Engelhardt, S., Zhu, D., Yuan, Y. (eds.) DGM4MICCAI 2022. LNCS, vol. 13609, pp. 117–126. Springer, Cham (2022). https://doi.org/10.1007/978-3-031-18576-2_12

23. Sajjadi, M.S., Bachem, O., Lucic, M., Bousquet, O., Gelly, S.: Assessing generative models via precision and recall. In: Advances in Neural Information Processing Systems, vol. 31 (2018)

24. Salimans, T., Goodfellow, I., Zaremba, W., Cheung, V., Radford, A., Chen, X.: Improved techniques for training GANs. In: Advances in Neural Information Processing Systems, vol. 29 (2016)

25. Sauer, A., Schwarz, K., Geiger, A.: StyleGAN-XL: scaling styleGAN to large diverse datasets. In: ACM SIGGRAPH 2022 Conference Proceedings, pp. 1–10 (2022)

26. Schaum, N., et al.: Ageing hallmarks exhibit organ-specific temporal signatures. Nature **583**(7817), 596–602 (2020)

27. Sudlow, C., et al.: UK biobank: an open access resource for identifying the causes of a wide range of complex diseases of middle and old age. PLoS Med. **12**(3), e1001779 (2015)

28. Sun, L., Chen, J., Xu, Y., Gong, M., Yu, K., Batmanghelich, K.: Hierarchical amortized GAN for 3d high resolution medical image synthesis. IEEE J. Biomed. Health Inform. **26**(8), 3966–3975 (2022)

29. Tsitsulin, A., et al.: The shape of data: intrinsic distance for data distributions. In: International Conference on Learning Representations (2020)

30. Vaswani, A., et al.: Attention is all you need. In: Guyon, I., et al. (eds.) Advances in Neural Information Processing Systems, vol. 30. Curran Associates, Inc. (2017)

31. Zhang, H., Goodfellow, I., Metaxas, D., Odena, A.: Self-attention generative adversarial networks. In: International Conference on Machine Learning, pp. 7354–7363. PMLR (2019)

32. Zhang, R., Isola, P., Efros, A.A., Shechtman, E., Wang, O.: The unreasonable effectiveness of deep features as a perceptual metric. In: 2018 IEEE/CVF Conference on Computer Vision and Pattern Recognition (CVPR), pp. 586–595. IEEE Computer Society (2018)

MIM-OOD: Generative Masked Image Modelling for Out-of-Distribution Detection in Medical Images

Sergio Naval Marimont[1]([✉]) [iD], Vasilis Siomos[1] [iD], and Giacomo Tarroni[1,2] [iD]

[1] CitAI Research Centre, City, University of London, London, UK
{sergio.naval-marimont,vasilis.siomos,giacomo.tarroni}@city.ac.uk
[2] BioMedIA, Imperial College London, London, UK

Abstract. Unsupervised Out-of-Distribution (OOD) detection consists in identifying anomalous regions in images leveraging only models trained on images of healthy anatomy. An established approach is to *tokenize* images and model the distribution of tokens with Auto-Regressive (AR) models. AR models are used to 1) identify anomalous tokens and 2) in-paint anomalous representations with in-distribution tokens. However, AR models are slow at inference time and prone to *error accumulation* issues which negatively affect OOD detection performance. Our novel method, MIM-OOD, overcomes both speed and error accumulation issues by replacing the AR model with two task-specific networks: 1) a transformer optimized to identify anomalous tokens and 2) a transformer optimized to in-paint anomalous tokens using masked image modelling (MIM). Our experiments with brain MRI anomalies show that MIM-OOD substantially outperforms AR models (DICE 0.458 vs 0.301) while achieving a nearly 25x speedup (9.5 s vs 244 s).

Keywords: out-of-distribution detection · unsupervised learning · masked image modelling

1 Introduction

Supervised deep learning approaches achieve state of the art performance in many medical image analysis tasks [10], but they require large amounts of manual annotations by medical experts. The manual annotation process is expensive and time-consuming, can lead to errors and suffers from inter-operator variability. Furthermore, supervised methods are specific to the anomalies annotated, which is in contrast with the diversity of naturally occurring anomalies. These limitations severely hinder applications of deep learning methods in the clinical practice.

Supplementary Information The online version contains supplementary material available at https://doi.org/10.1007/978-3-031-53767-7_4.

A. Mukhopadhyay et al. (Eds.): DGM4MICCAI 2023, LNCS 14533, pp. 35–44, 2024.
https://doi.org/10.1007/978-3-031-53767-7_4

Unsupervised Out-of-Distribution (OOD) detection methods propose to bypass these limitations by seeking to identify anomalies relying only on anomaly-free data. Generically, the so-called *normative* distribution of healthy anatomy is modelled by leveraging a training dataset of healthy images, and anomalous regions in test images are identified if they differ from the learnt distribution. A common strategy consists in modelling the *normative* distribution using generative models [1]. In particular, recent two-stage approaches have shown promising results [11,13,14,20]. In order to segment anomalies, a first stage encodes the image into discrete latent representations referred as *tokens*, from a VQ-VAE [12]. The second stage aims to model the likelihood of individual *tokens*, so a low likelihood can be used to identify those tokens that are not expected in the distribution of normal anatomies. Auto-Regressive (AR) modelling is the most common approach to model latent representation distributions [12]. Furthermore, once the anomalous *tokens* are identified, the generative capabilities of the AR model allow to in-paint anomalous regions with in-distribution *tokens*. By decoding the now in-distribution *tokens*, it is possible to obtain *healed* images, and anomalies can be localized with the pixel-wise residuals between original and *healed* images [11,13,14,20].

Although this strategy is effective, AR modelling requires defining an artificial order for the *tokens* so the latent distribution can be modelled as a sequence, and consequently suffer from two important drawbacks: 1) inference/image generation requires iterating through the latent variable sequence, which is computationally expensive, and 2) the fixed sequence order leads to error accumulation issues [13]. Error accumulation issues occur when AR models find OOD *tokens* early on the sequence and the *healed*/sampled sequence diverges from the original image, causing normal *tokens* to also be replaced.

Contributions: we propose MIM-OOD, a novel approach for OOD detection with generative models that overcomes the aforementioned issues to outperform equivalent AR models both in accuracy and speed. Our main contributions are:

- Instead of a single model for both tasks, MIM-OOD consists of two bidirectional Transformer networks: 1) the **Anomalous Token Detector**, trained to identify anomalous *tokens*, and 2) the **Generative Masked Visual Token Model (MVTM)**, trained using the masked image modelling (MIM) strategy and used to in-paint anomalous regions with in-distribution tokens. To our knowledge, it is the first time MIM is leveraged for this task.
- We evaluated MIM-OOD on brain MRIs, where it substantially outperformed AR models in detecting gliomas anomalies from BRATS dataset (DICE 0.458 vs 0.301) while also requiring a fraction of the inference time (9.5 s vs 244 s).

2   Related Works

Generative models have been at the core of the literature in unsupervised OOD detection in medical image analysis. Vanilla approaches using Variational Auto-Encoders (VAE) [8] are based on the assumption that VAEs trained on healthy data would not be able to reconstruct anomalous regions and consequently voxel-wise residuals could be used as Anomaly Score (AS) [1]. In [21], authors found that the KL-divergence term in the VAE loss function could be used to both detect anomalous samples and localise anomalies in pixel space using gradient ascent. Chen et al. [5] suggested using the gradients of the VAE loss function w.r.t. pixels values to iteratively *heal* the images and turn them into in-distribution samples, in a so-called *restoration* process. Approaches using Generative Adversarial Networks (GANs) [7] assume that anomalous samples are not encoded in the normative distribution and that AS can be derived from pixel-wise residuals between test samples and reconstructions [16]. As introduced previously, the most recent approaches based on generative models rely on two-stage image modelling [11,14,20]: the first stage is generally a *tokenizer* [12] that encodes images in discrete latent representations (referred as *tokens*), followed by a second stage that learns the *normative* distribution of *tokens* leveraging AR models. To overcome the limitations of AR modelling, Pinaya et al. [13] propose replacing AR with a latent diffusion model [15].

A novel and efficient strategy to model the latent distribution is generative masked image modelling (MIM), which has been used for image generation [4]. MIM consists in 1) dividing the variables of a multivariate joint probability distribution into masked and visible subsets, and in 2) modelling the probability of masked variables given visible ones. Masked Image Modelling strategy, when applied to latent visual *tokens* is referred to as Masked Visual Token Modelling (MVTM). MVTM produces high quality images by iteratively masking and resampling the latent variables where the model is less confident. In [9], authors improve the quality of the generated images by training an additional model, named *Token Critic*, to identify which latent variables require resampling by the MVTM model. The Token Critic is trained to identify which *tokens* are sampled from the model vs which tokens were in the original image. Token Critic and MVTM address the tasks of identifying inconsistent tokens and in-painting masks tokens, respectively, which are similar to the roles of AR models in the unsupervised OOD detection literature. We took inspiration from these efficient strategies to design our novel MIM-OOD method.

3   Method

3.1   Vector Quantized Variational Auto-encoders

Vector Quantized Variational Auto-Encoders (VQ-VAEs) [12] encode images $x \in \mathbb{R}^{H \times W}$ into representations $z_q \in \mathbf{K}^{H/f \times W/f}$ where $\mathbf{K}$ is a set of discrete representations, referred as *tokens*, and f is a downsampling factor measuring the spatial compression between pixels and *tokens*. A VQ-VAE first encodes the images to a continuous space $z_e \in \mathbb{R}^{D \times H/f \times W/f}$. Continuous representations are then discretized using an embedding space with $|\mathbf{K}|$ embeddings $e_k \in \mathbb{R}^D$. Specifically, *tokens* are defined as the index of the embedding vector e_k nearest to z_e: $z_q = argmin_j \|z_e - e_j\|_2$. For a detailed explanation, please refer to the original paper [12].

3.2   Generative Masked Visual Token Modelling (MVTM)

Masked modelling consists in learning the probability distribution of a set of occluded variables based on a set of observed ones from a given multivariate distribution. Occlusions are produced by replacing the values of variables with a special $[MASK]$ *token*. Consequently, the task is to learn $p(y_M \mid y_U)$, where y_M, y_U are the masked and unmasked exclusive subsets of Y. The binary mask $\mathbf{m} = [m_i]_{i=1}^{N}$ defines which variables are masked: if $i \in M$ then $m_i = 1$ and y_i is replaced with $[MASK]$. The training objective is to maximize the marginal cross-entropy for each masked *token*:

$$\mathcal{L}_{MVTM} = \sum_{\forall y_i \in Y_M} \log p_\phi(y_i \mid Y_U) \tag{1}$$

In MVTM, Y is the set of *tokens* $z_q \in \mathbf{K}^{H/f \times W/f}$ encoding image x. We use a multi-layer bi-directional Transformer to model the probability $p(y_i \mid Y_U)$ given the masked input. We optimize Transformer weights ϕ using back-propagation to minimize the cross-entropy between the ground-truth *tokens* and predicted *token* for the masked variables. The upper section in Fig. 1 describes the MVTM task.

The MVTM model can be leveraged to both in-paint latent regions using its generative capabilities and to identify anomalous *tokens*. A naive approach to identify anomalous *tokens* would be to use the predicted $p(y_i \mid Y_U)$ to identify *tokens* with low likelihood. However, by optimising over the marginals for each masked *token*, the model learns the distribution of each of the masked variables independently and fails to model the *joint* distribution of masked *tokens* [9]. Given the above limitation we introduce a second latent model specialized to identify anomalous *tokens*.

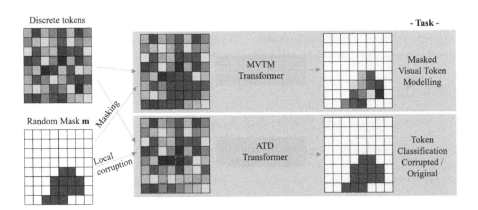

Fig. 1. Training tasks diagram. Given a random mask **m** and an healthy image representation (generated by the VQ-VAE), two Transformers are trained. Upper: MVTM is trained to predict masked *tokens*. Lower: Anomalous Token Detector (ATD) is trained to identify locally corrupted *tokens*.

3.3   Anomalous Token Detector (ATD)

The Anomalous Token Detector (ATD) is trained to identify local corruptions, similar to the task introduced in [17] but in latent space instead of pixel-space. We create the corruptions by replacing the *tokens* in mask $\mathbf{m}$ with random *tokens*. Consequently, our ATD bi-directional Transformer θ receives as input $\tilde{y} = y \odot (1-m) + r \odot m$, where r is a random set of *tokens* and $\odot$ is element-wise multiplication. It is trained to minimize a binary cross-entropy objective:

$$\mathcal{L}_{TC} = \sum_{i}^{N} m_i \log p_\theta(y_i) + (1 - m_i) \log(1 - p_\theta(y_i)) \tag{2}$$

The bottom section in Fig. 1 describes the proposed ATD task. Both the MVTM and ATD masks were generated by a random walk of a brush with a randomly changing width.

3.4   Image Restoration Procedure

At inference time, our goal is to evaluate if an image is consistent with the learnt healthy distribution and, if not, heal it by applying local transformations that replace anomalies with healthy tissue to generate a *restoration*. We can then localise anomalies by comparing the restored and original images. The complete MIM-OOD pipeline consists of the following steps:

1. **Tokenise** the image using the VQ-VAE Encoder.
2. **Identify anomalous tokens** by selecting *tokens* with an ATD prediction score greater than a threshold λ. Supplementary Materials include validation set results for different λ values.
3. **Restore anomalous tokens** by in-painting anomalous *tokens* with generative masked modelling. We sample tokens based on the likelihood assigned by MVTM: $\hat{y}_t \sim p_\phi(y_t \mid Y_U)$ and replace original tokens with samples in masked positions: $y_{t+1} = y_t \odot (1 - m) + \hat{y}_t \odot m$.
4. **Decode** *tokens* using the VQ-VAE Decoder to generate the restoration x_T.
5. **Compute the Anomaly Score (AS)** as the pixel-wise residuals between restoration x_T and original image x_0: $AS = |x_T - x_0|$. AS is smoothed to remove edges with min and $average - pooling$ filters [11].

Note that sampling from MVTM is done independently for each *token*, so it is possible that sampled *tokens* are inconsistent with each other. To address this issue, we can iterate T times steps 2 and 3. Additionally, we can also generate R multiple restorations per input image (i.e., repeating for each restoration step 3). We evaluated different values of T steps and R restorations, and validation set results are included in the Supplementary Materials. Figure 2 describes the end-to-end inference pipeline.

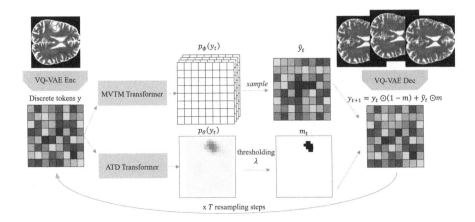

Fig. 2. MIM-OOD pipeline: the Anomalous Token Detector predicts the likelihood of *tokens* being anomalous. *Tokens* with ATD prediction score $> \lambda$ are deemed anomalous and replaced with samples from the MVTM model. We perform T iterations of the above procedure in parallel to generate R number of restorations.

4   Experiments and Results

We evaluated MIM-OOD on brain MRIs (training on a dataset of normal images and testing on one with anomalies) and compared our approach to an AR benchmark thoroughly evaluated in the literature [11,14]. Two publicly available datasets were used:

- The Human Connectome Project Young Adult (HCP) dataset [18] with images of 1,113 young and healthy subjects which we split into a training set with 1013 images and a validation set with 100 images.
- The Multimodal Brain Tumor Image Segmentation Benchmark (BRATS) [2], from the 2021 challenge. The dataset contains images with gliomas and comes with ground-truth segmentation masks highlighting their location. We randomly selected 100 images as validation set and 200 images as test set.

From both datasets we obtained pre-processed, skull-stripped T2-weighted structural images. We re-sampled both datasets to a common isotropic spacing of 1 mm and obtained random axial slices of 160×160 pixels. Intensity values were clipped to percentile 98 and normalized to the range [0, 1]. Latent models were only trained with random flip and rotations augmentations. We included broader-than-usual intensity augmentations during VQ-VAE training.

Network Architecture and Implementation Details: For the VQ-VAE we used the architecture from [6] and trained with MAE reconstruction loss for 300k steps. We used a codebook with $|\mathbf{K}| = 256$ and $D = 256$ and a downsampling factor $f = 8$. For the AR, ATD and MVTM we used a vanilla Transformer [19] with depth 12, layer normalization, and a stochastic depth of 0.1. Both the ATD and the MVTM Transformers are bi-directional since there is no sequence order to be enforced, in contrast to AR models. Transformers were trained for 200k steps.

We used AdamW with a cosine annealing scheduler, starting with learning rate of 5×10^{-5} and weight decay 1×10^{-5}. Our MONAI [3] implementation and trained models are made publicly available in[1].

Performance Evaluation: The role of the latent models (AR and MVTM) is both to identify anomalous latent variables and to replace them with in-distribution values for restoration. To evaluate the identification task, we setup a proxy task of classifying as anomalous *tokens* corresponding to image areas where anomalies are present. To this end, we downsample annotated ground-truth labels by the VQ-VAE scaling factor $f = 8$. Using *token* likelihood as a Anomaly Score (AS) in latent space we computed the Average Precision (AP) and the Area Under the Receiver Operating Characteristic curve (AUROC). Table 1 summarizes the proxy task results on the validation set. Our proposed approach relying on the ATD substantially outperforms the competitors.

Table 1. Results for Anomalous Token identification in BRATS validation set.

Method	AP	AUROC
AR [11]	0.054	0.773
MVTM [4]	0.084	0.827
MVTM + Token Critic [9]	0.041	0.701
MIM-OOD (Ours)	**0.186**	**0.859**

We then evaluate MIM-OOD's capability to localise anomalies in image space on the test set. We report the best achievable $[DICE]$ score following the conventions in recent literature [1,14]. Additionally, we report AP, AUROC and inference time per batch of 32 images (IT (s)) using a single Nvidia RTX3090.

Table 2. Results for Pixel-wise Anomaly Detection in BRATS test set.

Method	[DICE]	AP	AUROC	IT (s)
AR restoration $R = 4$ [(1)] [11]	0.301	0.191	0.891	244
MVTM + Token Critic R = 4 [9]	0.201	0.131	0.797	5.4
MIM-OOD R = 4 (Ours) [(2)]	0.458	0.399	0.926	9.5
MIM-OOD R = 8 (Ours) [(2)]	**0.461**	**0.404**	**0.928**	19.9

1 - Implementation from [11] with 4 restorations and $\lambda_{NLL} = 6$, using Transformer instead of PixelSNAIL AR architecture.
2 - Our approach with 4 and 8 restorations respectively, $\lambda = 0.005$ and $T = 8$.

---

[1] https://github.com/snavalm/MIM-OOD.

The results in Table 2 show that MIM-OOD improves the [*DICE*] score by 15 points (0.301 vs 0.458) when using the same number of restorations ($R = 4$), while at the same time reducing the inference time (244 s vs 9.5 s for a batch of 32 images). These results are consistent with the previous anomalous *token* identification task where we showed that our ATD model was able to better identify anomalous *tokens* requiring restorations. Qualitative results are included in Fig. 3 and in Supplementary Materials.

It is worth noting that our AR baseline slightly underperforms compared to previous published results (DICE of 0.301 vs 0.328 in BRATS dataset [13, 14]) due to different experimental setups: different modalities (T2 vs FLAIR), training sets (HCP with N = 1,013 vs UK Biobank with N = 14,000) and pre-processing pipelines. While both T2 and FLAIR modalities are common in the OOD literature, T2 was chosen for our experiments because of the availability of freely accessible anomaly-free datasets. The differences in experimental setups makes results not directly comparable, however our approach shows promising performance and high efficiency when comparing with both AR Ensemble ([14] DICE 0.537 and 4,907 s per 100 image batch) and Latent Diffusion ([13] DICE 0.469 and 324 s per 100 image batch).

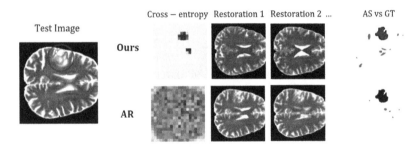

Fig. 3. Qualitative comparison. Our method outperforms the AR at both identifying the anomalous latent *tokens* and generating restorations where the lesion has been healed. This translates to a more reliable AS map (pink) vs ground truth (black). (Color figure online)

5   Conclusion

We developed MIM-OOD, a novel unsupervised anomaly detection technique that, to our knowledge, leverages for the first time the concept of Generative Masked Modelling for this task. In addition, we introduce ATD, a novel approach to identify anomalous latent variables which synergizes with the MVTM to generate healed restorations. Our results show that our technique outperforms previous AR-based approaches in unsupervised glioma segmentation in brain MRI. In the future, we will perform further evaluations of our approach, testing it on data with other brain pathologies and using additional image modalities.

References

1. Baur, C., Denner, S., Wiestler, B., Albarqouni, S., Navab, N.: Autoencoders for unsupervised anomaly segmentation in brain MR images: a comparative study. Med. Image Anal. **69**, 101952 (2021)
2. Menze, B.H., et al.: The multimodal brain tumor image segmentation benchmark (BRATS). IEEE Trans. Med. Imaging **34**(10), 1993–2024 (2014). https://doi.org/10.1109/TMI.2014.2377694. Epub 2014 Dec 4
3. Cardoso, M.J., et al.: MONAI: an open-source framework for deep learning in healthcare, November 2022. arXiv:2211.02701 [cs]
4. Chang, H., Zhang, H., Jiang, L., Liu, C., Freeman, W.T.: MaskGIT: masked generative image transformer. In: Proceedings of the IEEE/CVF Conference on Computer Vision and Pattern Recognition, pp. 11315–11325 (2022)
5. Chen, X., You, S., Tezcan, K.C., Konukoglu, E.: Unsupervised lesion detection via image restoration with a normative prior. In: Proceedings of The 2nd International Conference on Medical Imaging with Deep Learning PMLR, vol. 102, pp. 540–556 (2020)
6. Esser, P., Rombach, R., Ommer, B.: Taming transformers for high-resolution image synthesis. In: Proceedings of the IEEE/CVF Conference on Computer Vision and Pattern Recognition, pp. 12873–12883 (2021)
7. Goodfellow, I.J., et al.: Generative adversarial networks. In: Advances in Neural Information Processing Systems, pp. 2672–2680 (2014)
8. Kingma, D.P., Welling, M.: Auto-encoding variational bayes. In: The 2nd International Conference on Learning Representations (ICLR) (2013)
9. Lezama, J., Chang, H., Jiang, L., Essa, I.: Improved masked image generation with token-critic. In: Avidan, S., Brostow, G., Cissé, M., Farinella, G.M., Hassner, T. (eds.) ECCV 2022, Part XXIII. LNCS, vol. 13683, pp. 70–86. Springer, Cham (2022). https://doi.org/10.1007/978-3-031-20050-2_5
10. Litjens, G., et al.: A survey on deep learning in medical image analysis. Med. Image Anal. **42**, 60–88 (2017)
11. Naval Marimont, S., Tarroni, G.: Anomaly detection through latent space restoration using vector quantized variational autoencoders. In: 2021 IEEE 18th International Symposium on Biomedical Imaging (ISBI), pp. 1764–1767. IEEE (2021)
12. van den Oord, A., Vinyals, O., Kavukcuoglu, K.: Neural discrete representation learning. In: NIPS 2017: Proceedings of the 31st International Conference on Neural Information Processing Systems, pp. 6309–6318 (2017)
13. Pinaya, W.H.L., et al.: Fast unsupervised brain anomaly detection and segmentation with diffusion models. arXiv preprint arXiv:2206.03461 (2022)
14. Pinaya, W.H.L., et al.: Unsupervised brain imaging 3D anomaly detection and segmentation with transformers. Med. Image Anal. **79**, 102475 (2022)
15. Rombach, R., Blattmann, A., Lorenz, D., Esser, P., Ommer, B.: High-resolution image synthesis with latent diffusion models. In: Proceedings of the IEEE/CVF Conference on Computer Vision and Pattern Recognition, pp. 10684–10695 (2022)
16. Schlegl, T., Seeböck, P., Waldstein, S.M., Langs, G., Schmidt-Erfurth, U.: f-AnoGAN: fast unsupervised anomaly detection with generative adversarial networks. Med. Image Anal. **54**, 30–44 (2019)
17. Tan, J., Hou, B., Batten, J., Qiu, H., Kainz, B.: Detecting outliers with foreign patch interpolation. Mach. Learn. Biomed. Imag. **1**, 1–27 (2022)
18. Van Essen, D.C., et al.: The Human ConnectomeProject: a data acquisition perspective. Neuroimage **62**(4), 2222–2231 (2012)

19. Vaswani, A., et al.: Attention is all you need. In: Advances in Neural Information Processing Systems, vol. 30 (2017)
20. Wang, L., Zhang, D., Guo, J., Han, Y.: Image anomaly detection using normal data only by latent space resampling. Appl. Sci. **10**(23), 8660 (2020)
21. Zimmerer, D., Isensee, F., Petersen, J., Kohl, S., Maier-Hein, K.: Unsupervised anomaly localization using variational auto-encoders. In: Shen, D., et al. (eds.) MICCAI 2019. LNCS, vol. 11767, pp. 289–297. Springer, Cham (2019). https://doi.org/10.1007/978-3-030-32251-9_32

Towards Generalised Neural Implicit Representations for Image Registration

Veronika A. Zimmer[1,2,3(✉)], Kerstin Hammernik[1,5], Vasiliki Sideri-Lampretsa[1], Wenqi Huang[1], Anna Reithmeir[1,2,4], Daniel Rueckert[1,3,5], and Julia A. Schnabel[1,2,4,6]

[1] School of Computation, Information and Technology, Technical University of Munich, Munich, Germany
{k.hammernik,veronika.zimmer}@tum.de
[2] Helmholtz Munich, Munich, Germany
[3] School of Medicine, Klinikum Rechts der Isar, Technical University of Munich, Munich, Germany
[4] Munich Center for Machine Learning (MCML), Munich, Germany
[5] Department of Computing, Imperial College London, London, UK
[6] King's College London, London, UK

Abstract. Neural implicit representations (NIRs) enable to generate and parametrize the transformation for image registration in a continuous way. By design, these representations are image-pair-specific, meaning that for each signal a new multi-layer perceptron has to be trained. In this work, we investigate for the first time the potential of existent NIR generalisation methods for image registration and propose novel methods for the registration of a group of image pairs using NIRs. To exploit the generalisation potential of NIRs, we encode the fixed and moving image volumes to latent representations, which are then used to condition or modulate the NIR. Using ablation studies on a 3D benchmark dataset, we show that our methods are able to generalise to a set of image pairs with a performance comparable to pairwise registration using NIRs when trained on $N = 10$ and $N = 120$ datasets. Our results demonstrate the potential of generalised NIRs for 3D deformable image registration.

Keywords: Image registration · Neural implicit representation · Generalisation · Periodic activation functions

1 Introduction

Image registration is the process of aligning two or more images taken from different sources or at different times, so that corresponding features in the images are in the same spatial/anatomical position. The registration of medical images is an important processing step for the reliable and quantitative interpretation of changes occurring in multiple images, e.g., for disease monitoring [11] or motion

---

V. A. Zimmer and K. Hammernik–Authors contributed equally.

A. Mukhopadhyay et al. (Eds.): DGM4MICCAI 2023, LNCS 14533, pp. 45–55, 2024.
https://doi.org/10.1007/978-3-031-53767-7_5

correction and estimation [6,8,19]. Classical registration methods rely on pairwise optimisation strategies where a distance term between images is minimized over a space of spatial transformations. Popular transformation models include B-Splines [15,19] and velocity and flow fields [2,4], among others. In recent years, several learning-based solutions have been proposed [10], where the most popular ones learn to predict the spatial transformations as displacement fields [3], diffeomorphisms [5,18] or as parameters of the transformation model [7] between the images using convolutional neural networks (CNNs). These methods leverage the information of multiple image pairs during training and predict the transformation between unseen images as the output of the CNN at inference.

Recently, neural implicit representations (NIRs) have been proposed to encode continuous signals such as images in the weights of a multi-layer perceptron (MLP) [16,21,23]. The spatial coordinates of the image grid are fed into the MLP which is trained to approximate the image intensities. One advantage is that this representation is continuous: the signal can be sampled at any point in space without the need for interpolation. The expressiveness of these continuous representations has been enhanced by Fourier Encoding [23] or periodic activation functions [21]. The former work [23] proposed to map the input coordinates to a Fourier feature space to overcome the difficulties of the MLP to learn high-frequency functions. The latter work [21] achieves this by using sine activation functions in the MLP. With both strategies NIRs are successfully used to represent continuous functions in various applications in computer vision and graphics [17]. In the medical imaging domain, NIRs have been investigated for magnetic resonance image reconstruction in the image domain [20] and k-space domain [13], radiation therapy [25], or image segmentation [14].

NIRs have also been studied to parametrize the transformation in a pairwise image registration setting [26]. This is different from other learning-based approaches, where neural networks are used to predict a transformation, while here the authors propose to represent the transformation by a neural network. The inputs to the MLP are the coordinates of the image grid, while the outputs are the coordinate's displacements (instead of intensities). The MLP is trained to minimize the image distance for each image pair, and the weights encode the transformation between these images. One advantage of this implicit transformation representation is that it is continuous. A displacement can be estimated for any voxel coordinate in the image space without interpolation, making the method independent of the image grid resolution.

By design, a NIR is signal-specific and does not generalise to other signals. This means that a transformation-encoding MLP for image registration is image-pair-specific and for each new image pair, a new MLP has to be trained. This is similar to the pairwise optimisation strategies in classical image registration.

Previous works have proposed several approaches for generalisation of NIRs for continuous image representation, including hypernetworks [21,22], conditioned MLPs [1] and modulation of the periodic activation functions [16]. In [1], NIRs are used for shape representations and generalised by conditioning the MLP on a latent vector as an additional input to the MLP. The latent vector is trained simultaneously with the MLP from the latent code of an autoencoder. In [16], the authors propose to use the latent representations of local image patches to modulate the amplitude of the sine activation functions. They use two MLPs, a synthesis network using sine activations (the NIR) and a modulator network using ReLU activations. The modulator network produces outputs (using the latent codes) to modulate the amplitude of the sine activations of each layer in the synthesis network.

Generalisation would have several advantages for NIRs in image registration. First, it would not be necessary to train a new network for every new image pair, which might be inefficient for large datasets even when using lightweight networks such as MLPs. Second, the network could leverage the image features as additional inputs, which are so far only used in the loss function to train the MLP. Third, the training on multiple image pairs could yield image feature sharing across the dataset, potentially benefiting the registration performance.

In this work, we adopt the methodological framework from [26], and investigate how NIRs can be generalised for image registration. In particular, we explore how conditional MLPs and the modulation of periodic activation functions can be leveraged to train a NIR for a group of images. We utilize global image latent codes to inject the information of the images into the NIR. In [16], local latent codes from image patches were used, but this approach is unfeasible for 3D image registration, as for each spatial coordinate an image patch would have to be sampled. This work presents the first step towards full generalisation of NIR for image registration. Our contributions are three-fold:

- We explore for the first time generalisation techniques for NIRs in 3D image registration.
- We propose novel strategies for generalisation of NIR adapted to the image registration setting.
- In an ablation study, we show that modulation of periodic activations functions, based on features extracted through a 3D encoder network, show high potential for generalisation of NIRs on benchmark registration data.

2   Methodology

Given two d-dimensional images, a fixed image F and a moving image M with $F, M : \Omega \subset \mathbb{R}^d \to \mathbb{R}$. The aim of image registration is to find an optimal spatial transformation $\phi : \mathbb{R}^d \to \mathbb{R}^d$ such that the transformed moving image is most similar to the fixed image $M \circ \phi \approx F$. Typically, this is formulated as an optimisation problem $\phi^* = \arg\max_\phi \mathcal{J}(F, M, \phi)$ where the distance between the images is minimised subject to constraints put on the transformation. We denote the objective function $\mathcal{J}$ as

$$\mathcal{J}(F, M, \phi) = \mathcal{D}(F, M \circ \phi) + \alpha \mathcal{R}(\phi), \qquad (1)$$

with an image distance measure $\mathcal{D}$, a regulariser $\mathcal{R}$ posing some additional constraints on the transformation ϕ and a regularisation parameter $\alpha \in \mathbb{R}^+$ determining the influence of $\mathcal{R}$ on the solution. In classical methods, this problem is

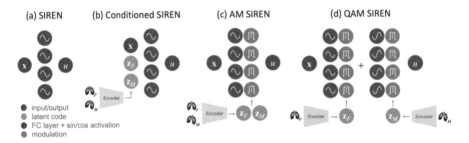

Fig. 1. Overview of architectures for neural implicit representations. (a) Image-pair-specific MLP with ReLU or SIREN (with sine activation functions). (b) Conditioned SIREN (MLP + sine activations). (c) Amplitude-Modulated SIREN. (d) Quadrature Amplitude-Modulated SIREN.

solved iteratively for a pair of images, e.g., using gradient-based optimisation. We now introduce NIRs for image registration and generalisation techniques.

2.1 Neural Implicit Representations for Image Registration

NIR can be used to model the transformation ϕ [26]. An MLP f_θ with trainable parameters θ is used to approximate the transformation $\phi(\mathbf{x}) = \mathbf{x} + u(\mathbf{x})$ with $\mathbf{x} \in \Omega$ between a given pair of images F, M by estimating $u(\mathbf{x}) = f_\theta(\mathbf{x})$. The L-layer network is modelled as $f_\theta = f_L \circ f_{L-1} \circ \ldots \circ f_1$, where

$$\mathbf{h}_l = f_l(\mathbf{h}_{l-1}) = \psi(W_l \mathbf{h}_{l-1} + b_l), 0 < l \leq L.$$

The dense matrix W_l and the bias b_l are learned, while the activation function ψ is fixed to either ReLU or sine activations (SIRENS). The variable $\mathbf{h}_l$ denotes the hidden feature vector for the l-th layer, with the initial feature vector $\mathbf{h}_0 = \mathbf{x}$.

2.2 Generalised Neural Implicit Representations

Our proposed generalised NIRs are based on conditioned MLPs [1] and modulated SIRENs [16]. We adapt both approaches and describe novel strategies to generalise NIRs to image registration. All presented approaches have in common that we first obtain individual latent representations $\mathbf{z}_F$ and $\mathbf{z}_M$ for the fixed and moving image, respectively which are processed in three different ways.

Conditioned Networks. In conditioned networks, the latent feature representation $\mathbf{z}_F$ and $\mathbf{z}_M$ are appended to the input coordinates $\mathbf{x}$, yielding $\mathbf{h}_0 = [\mathbf{x}, \mathbf{z}_F, \mathbf{z}_M]$. We use a SIREN network as base network, i.e., ψ equals the sine activation. Hence, we refer to this architecture as *conditioned SIREN* or *C SIREN*.

Amplitude-Modulated (AM) Periodic Activation Functions. While conditioned SIRENs use $\mathbf{z}_F$ and $\mathbf{z}_F$ only at the input stage, the latent representations $\mathbf{z}_F$ and $\mathbf{z}_M$ are used to modulate the amplitude of the sine activations ψ in every layer. The modulation is realized by a vector α_l following

$$\alpha_l = \sigma(W_{\alpha,l}[\alpha_{l-1}, \mathbf{z}_F, \mathbf{z}_M]^\top + b_{\alpha,l}), \quad 0 < l \le L \tag{2}$$
$$\alpha_0 = \sigma(W_{\alpha,l}[\mathbf{z}_F, \mathbf{z}_M]^\top + b_{\alpha,l}) \tag{3}$$

where $W_{\alpha,l}$ is a dense matrix, $b_{\alpha,l}$ denotes the bias of the MLP. The modulated SIREN reads as point-wise product between the modulation vector α_l and the hidden features h_l, denoted by $\odot$

$$\mathbf{h}_l = \alpha_l \odot f_l(\mathbf{h}_{l-1}) = \alpha_l \odot \psi(W_l \mathbf{h}_{l-1} + b_l), \quad 0 < l \le L.$$

This modulations implicitly influences the frequency of subsequent layers [16]. We refer to this modulation as *AM SIREN*.

Quadrature Amplitude-Modulated (QAM) Periodic Activation Functions. Up to know, we stacked the latent codes $\mathbf{z}_F$ and $\mathbf{z}_M$ for further processing. However, we can also process them separately. Inspired by quadrature amplitude modulation used in telecommunication systems, we propose to compose two modulated signal, which are then further processed by subsequent layers as

$$\mathbf{h}_l = \alpha_l \odot f_{1,l}(\mathbf{h}_{l-1}) + \beta_l \odot f_{2,l}(\mathbf{h}_{l-1})$$
$$= \alpha_l \odot \psi_1(W_{1,l}\mathbf{h}_{l-1} + b_{1,l}) + \beta_l \odot \psi_2(W_{2,l}\mathbf{h}_{l-1} + b_{2,l}), \quad 0 < l \le L.$$

The parameters $W_{1,l}$, $W_{2,l}$, $b_{1,l}$ and $b_{2,l}$ denote the dense matrices and biases of the two components. ψ_1 and ψ_2 are defined as the sine and cosine activation functions, respectively. The modulation parameters α and β are defined according to Eqs. (2) and (3). We denote this modulation technique as *QAM SIREN*.

2.3   Image Encoding Using Denoising Autoencoders

Image information is provided to the MLP in form of a latent vector $\mathbf{z} \in \mathbb{R}^Z$, encoding the features of the image in a low dimension space with dimension Z (see Fig. 1). We use a CNN encoder to encode the images F and M into latent vectors $\mathbf{z}_F, \mathbf{z}_M \in \mathbb{R}^Z$. The CNN has four layers with channel dimensions $[32, 64, 128, 256]$. Each layer consists of a convolutional layer with stride 2 followed by batch normalization, leaky ReLU activation and max pooling. The last layer is followed by a linear layer.

We pre-train the weights of the encoder as part of a denoising autoencoder. The decoder starts with a linear layer, followed by four layers consisting of transposed convolutions, batch normalization and leaky ReLU activation function.

3   Materials and Experiments

3.1   Data

We evaluate the performance of our methods on publicly available benchmark data from the Learn2Reg challenge 2022[1] [12]. The dataset consists of computed tomography (CT) images of the lung from 150 patients. The data is originally taken from the National Lung Screening Trial[2] (NLST) [24], a randomized multi-center study for screening and early detection of lung cancer. For each patient, images at inspiration and expiration are available and the goal of image registration is to estimate respiratory motion, which is important for multiple clinical tasks such as radiotherapy planning, and general assessment of the lungs.

For the Learn2Reg challenge, a subset of the data was pre-processed: all images were resampled to an isotropic resolution of 1.5 mm and cropped to an image size of [224, 192, 224]. The lungs were automatically segmented and keypoints (only for the training data) were automatically extracted. For the details of the pre-processing, lung segmentation and keypoint extraction, we refer to the Learn2Reg challenge 2022.

3.2   Experimental Design

We trained all methods using Eq. (1) with Normalized Cross Correlation as image distance measure $\mathcal{D}$ and Bending Energy [19] as regularization with $\alpha = 10$ as suggested in [26]. We designed different experiments to evaluate our proposed methods regarding different aspects: (i) generalisation strategy, (ii) MLP architecture, (iii) latent feature vector training, (iv) training dataset size.

First, we perform an ablation study on the different settings using a dataset of ten CT image pairs. As a baseline, we registered each pair using the NIR method (MLP and SIREN) by [26] using their publicly available code with the default parameters. We compare this pairwise approach to C SIRENs, AM SIRENs and QAM SIRENs (cp. Sect. 2.2), trained and tested on the same ten image pairs. We pre-trained the autoencoder using a denoising task on an independent set of images using a 64-dimensional latent vector. In the ablation study we test different architectures for the MLPs, in particular the number of layers (3 and 5 layers) and hidden units (256 and 512). In addition, we test the effect of freezing or unfreezing the weights of the image feature encoder. By unfreezing the weights, the encoder is fine-tuned simultaneously with the training of the MLP. Second, we evaluate the effect of a larger training set of $N = 120$ on the performance on the same ten datasets as before for AM SIREN with varying architectures.

All networks were trained using the ADAM optimiser with a learning rate of 10^{-5} and coordinate batch size of 10,000 until convergence. We implemented the models in PyTorch (1.13.1). The code is publicly available[3].

---

[1] https://learn2reg.grand-challenge.org/.

[2] https://www.cancer.gov/types/lung/research/nlst.

[3] https://github.com/vamzimmer/generalized_idir.

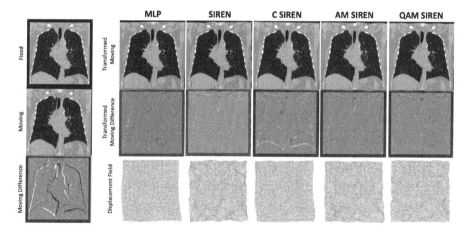

Fig. 2. Qualitative registration results using different pairwise (MLP, SIREN) methods and methods generalised over ten image pairs (C SIREN, AM SIREN, QAM SIREN).

3.3   Evaluation Measures

As ground truth deformations are typically not known, we evaluate the registration performance using the Dice overlap and the robust (95%) Hausdorff distance (HD95) on the lung masks and the mean squared error between the keypoints (MSE_{kp}). Additionally, we evaluate the plausibility of the resulting transformations by computing the standard deviation of the logarithm of the jacobian determinant of the displacement field (SDLogJ). We test for statistical significance using a paired Wilcoxon signed-rank test between the pairwise SIREN baseline and each generalized model. We consider significance at $p < 0.05$.

4   Results

The results for the ablation study for the generalisation over a set of ten image pairs are reported in Table 1 and qualitative results are shown in Fig. 2. AM SIREN yield competitive results to the baseline pairwise registration with SIREN, indicating that the generalisation on these ten image pairs was successful. We observe that with deeper MLPs and with fine-tuning the latent code (Enc) the performance increases, except for QAM SIREN. C Siren benefits most from latent code fine-tuning. However, at the same time the number of network weights increases up to 80 Mio, when a full 3D autoencoder is trained simultaneously with the (originally lightweight) MLP. The best results among the generalisation methods are obtained by AM SIREN with 5 layers and 256 hidden units (best overall HD95) and 512 hidden units (best overall Dice). For the latter model, all metrics do not show any statistical difference to the baseline SIREN. The results are similar to the pairwise SIREN with only 3 layers and 256 hidden units, indicating that deeper, more complex NIRs are necessary to

Table 1. Architecture Ablation. The generalisation of NIR for image registration to a group of image pairs ($N = 10$) is compared for several architectures (varying layers and hidden units) to the baseline approach of pairwise MLP and SIREN [26]. Different generalisation methods are C SIREN, AM SIREN and QAM SIREN either with fine-tuning of the latent image code (Enc) or without. Gray boxes: $p > 0.05$ (**not sig.**).

Method	Arch.	Parameters	MSE_{kp}	Dice	HD95
Initial	–	–	6.97 ± 2.21	0.928 ± 0.01	5.61 ± 2.17
C SIREN	3l-256	182,659	2.46 ± 0.93	0.966 ± 0.01	2.49 ± 0.99
	3l-256 Enc	80,181,444	2.00 ± 0.63	0.970 ± 0.01	2.63 ± 1.52
	3l-512	610,947	2.48 ± 1.00	0.969 ± 0.01	2.38 ± 0.96
AM SIREN	3l-256	363,523	1.51 ± 0.85	0.980 ± 0.01	1.72 ± 1.21
	3l-256 Enc	80,362,308	1.47 ± 0.67	0.981 ± 0.01	1.79 ± 1.27
	5l-256	692,227	1.34 ± 0.69	0.983 ± 0.01	$\mathbf{1.18 \pm 0.32}$
	3l-512	1,251,331	1.37 ± 0.86	0.983 ± 0.01	1.52 ± 1.17
	5l-512	2,433,027	1.24 ± 0.62	$\mathbf{0.984 \pm 0.01}$	1.24 ± 0.60
QAM SIREN	3l-256	627,971	1.78 ± 1.04	0.980 ± 0.01	1.59 ± 1.08
	3l-256 Enc	80,626,756	1.92 ± 1.22	0.978 ± 0.01	1.80 ± 1.28
	3l-512	2,304,515	1.44 ± 0.84	0.983 ± 0.00	1.20 ± 0.60
Pairwise MLP	3l-256	133,379	1.65 ± 0.81	0.974 ± 0.01	3.00 ± 3.84
Pairwise SIREN	3l-256	133,379	$\mathbf{1.13 \pm 0.48}$	$\mathbf{0.984 \pm 0.02}$	1.32 ± 0.50

Table 2. Generalisation of NIR on the same $N = 10$ image pairs as in Table 1. Training was performed with $N = 120$ datasets for Amplitude-Modulated (AM) SIRENs.

Method	Arch	MSE_{kp}	Dice	HD95
AM SIREN	3l-256	2.94 ± 0.98	0.958 ± 0.02	4.16 ± 2.27
	5l-256	2.24 ± 0.84	0.968 ± 0.01	2.75 ± 1.24
	3l-512	2.45 ± 0.85	0.966 ± 0.01	3.19 ± 1.51
	5l-512	$\mathbf{1.87 \pm 0.80}$	$\mathbf{0.973 \pm 0.01}$	$\mathbf{2.43 \pm 1.70}$

encode the transformations between multiple image pairs. QAM and C SIREN perform slightly worse, however, they still reach an MSE_{kp} of 2.48 as maximum. QAM SIREN outperforms C SIREN, emphasising the capabilities of modulated SIRENs for generalisation. For all methods, the SDLogJ is in the order of 10^{-4} with 0% of negative Jacobian determinants (min: 3.98; max: 4.03), indicating that the regularization correctly constrains the transformations.

The results for training on $N = 120$ for AM SIREN with varying architectures are reported in Table 2. Although the MLP was trained on many image pairs, the registration performance on the ten test pairs does not deteriorate, with a minimum MSE_{kp} of 1.87. This indicates that the AM SIREN can encode the transformation between multiple image pairs accurately. As expected, deeper MLPs perform better, as they can capture more information.

5 Discussion and Conclusion

In this work, we explored the generalisation abilities of NIRs for image registration. We investigated two approaches for generalising NIRs: conditioning the MLP with image feature vectors and modulating the periodic activation functions using the images encoder as a modulator network. Our experiments show the superior performance of the latter approach using modulation of activation.

To the best of our knowledge, this is the first work on generalising NIRs for image registration. To this end, we focused on generalising the registration to a group of image pairs and not yet to unseen images. We tested generalisation strategies which have been developed for signal approximation, but image registration is a much more complex task, and our results suggest that more research is needed to obtain complete generalisation. Our work presents a first step towards generalisation and shows potential directions of further research, especially with regards to the modulation of sine activation functions.

Another limitation of our study is that we present results from a single mono-modal benchmark dataset. In the future, we plan to include more diverse data and datasets, including multi-modal data. We also did not compare to other baseline registration methods such as [11,15] for classical pairwise and [3,9,18] for learning-based methods. Our objective was not to propose a new state-of-the-art but rather show a proof of concept for the generalisation of NIRs for image registration.

References

1. Amiranashvili, T., Lüdke, D., Li, H.B., Menze, B., Zachow, S.: Learning shape reconstruction from sparse measurements with neural implicit functions. In: International Conference on Medical Imaging with Deep Learning, pp. 22–34. PMLR (2022)
2. Ashburner, J.: A fast diffeomorphic image registration algorithm. Neuroimage **38**(1), 95–113 (2007)
3. Balakrishnan, G., Zhao, A., Sabuncu, M.R., Guttag, J., Dalca, A.V.: Voxelmorph: a learning framework for deformable medical image registration. IEEE Trans. Med. Imaging **38**(8), 1788–1800 (2019)
4. Beg, M.F., Miller, M.I., Trouvé, A., Younes, L.: Computing large deformation metric mappings via geodesic flows of diffeomorphisms. Int. J. Comput. Vision **61**, 139–157 (2005)
5. Dalca, A.V., Balakrishnan, G., Guttag, J., Sabuncu, M.R.: Unsupervised learning for fast probabilistic diffeomorphic registration. In: Frangi, A.F., Schnabel, J.A., Davatzikos, C., Alberola-López, C., Fichtinger, G. (eds.) MICCAI 2018. LNCS, vol. 11070, pp. 729–738. Springer, Cham (2018). https://doi.org/10.1007/978-3-030-00928-1_82

6. De Craene, M., et al.: Temporal diffeomorphic free-form deformation: application to motion and strain estimation from 3d echocardiography. Med. Image Anal. **16**(2), 427–450 (2012)

7. De Vos, B.D., Berendsen, F.F., Viergever, M.A., Sokooti, H., Staring, M., Išgum, I.: A deep learning framework for unsupervised affine and deformable image registration. Med. Image Anal. **52**, 128–143 (2019)

8. Gigengack, F., Ruthotto, L., Burger, M., Wolters, C.H., Jiang, X., Schafers, K.P.: Motion correction in dual gated cardiac pet using mass-preserving image registration. IEEE Trans. Med. Imaging **31**(3), 698–712 (2011)

9. Hansen, L., Heinrich, M.P.: Graphregnet: deep graph regularisation networks on sparse keypoints for dense registration of 3d lung CTS. IEEE Trans. Med. Imaging **40**(9), 2246–2257 (2021)

10. Haskins, G., Kruger, U., Yan, P.: Deep learning in medical image registration: a survey. Mach. Vis. Appl. **31**, 1–18 (2020)

11. Heinrich, M.P., Jenkinson, M., Brady, M., Schnabel, J.A.: MRF-based deformable registration and ventilation estimation of lung CT. IEEE Trans. Med. Imaging **32**(7), 1239–1248 (2013)

12. Hering, A., et al.: Learn2reg: comprehensive multi-task medical image registration challenge, dataset and evaluation in the era of deep learning. IEEE Trans. Med. Imaging (2022)

13. Huang, W., Li, H.B., Pan, J., Cruz, G., Rueckert, D., Hammernik, K.: Neural Implicit k-Space for Binning-Free Non-Cartesian Cardiac MR Imaging. In: Frangi, A., de Bruijne, M., Wassermann, D., Navab, N. (eds.) Information Processing in Medical Imaging, IPMI 2023, LNCS, vol. 13939, pp. 548–560. Springer, Cham (2023). https://doi.org/10.1007/978-3-031-34048-2_42

14. Khan, M.O., Fang, Y.: Implicit neural representations for medical imaging segmentation. In: Wang, L., Dou, Q., Fletcher, P.T., Speidel, S., Li, S. (eds.) Medical Image Computing and Computer Assisted Intervention, MICCAI 2022, MICCAI 2022, LNCS, vol. 13435, pp. 433–443. Springer, Cham (2022). https://doi.org/10.1007/978-3-031-16443-9_42

15. Klein, S., Staring, M., Murphy, K., Viergever, M.A., Pluim, J.P.: Elastix: a toolbox for intensity-based medical image registration. IEEE Trans. Med. Imaging **29**(1), 196–205 (2009)

16. Mehta, I., Gharbi, M., Barnes, C., Shechtman, E., Ramamoorthi, R., Chandraker, M.: Modulated periodic activations for generalizable local functional representations. In: Proceedings of the IEEE/CVF International Conference on Computer Vision, pp. 14214–14223 (2021)

17. Mildenhall, B., Srinivasan, P.P., Tancik, M., Barron, J.T., Ramamoorthi, R., Ng, R.: Nerf: representing scenes as neural radiance fields for view synthesis. Commun. ACM **65**(1), 99–106 (2021)

18. Mok, T.C.W., Chung, A.C.S.: Large deformation diffeomorphic image registration with Laplacian pyramid networks. In: Martel, A.L., et al. (eds.) MICCAI 2020, Part III. LNCS, vol. 12263, pp. 211–221. Springer, Cham (2020). https://doi.org/10.1007/978-3-030-59716-0_21

19. Rueckert, D., Sonoda, L.I., Hayes, C., Hill, D.L., Leach, M.O., Hawkes, D.J.: Non-rigid registration using free-form deformations: application to breast MR images. IEEE Trans. Med. Imaging **18**(8), 712–721 (1999)

20. Shen, L., Pauly, J., Xing, L.: NeRP: implicit neural representation learning with prior embedding for sparsely sampled image reconstruction. IEEE Trans. Neural Netw. Learn. Syst. (2022)

21. Sitzmann, V., Martel, J., Bergman, A., Lindell, D., Wetzstein, G.: Implicit neural representations with periodic activation functions. Adv. Neural. Inf. Process. Syst. **33**, 7462–7473 (2020)
22. Sitzmann, V., Rezchikov, S., Freeman, B., Tenenbaum, J., Durand, F.: Light field networks: neural scene representations with single-evaluation rendering. Adv. Neural. Inf. Process. Syst. **34**, 19313–19325 (2021)
23. Tancik, M., Srinivasan, P., Mildenhall, B., Fridovich-Keil, S., Raghavan, N., Singhal, U., Ramamoorthi, R., Barron, J., Ng, R.: Fourier features let networks learn high frequency functions in low dimensional domains. Adv. Neural. Inf. Process. Syst. **33**, 7537–7547 (2020)
24. Team, N.L.S.T.R.: The national lung screening trial: overview and study design. Radiology **258**(1), 243–253 (2011)
25. Vasudevan, V., et al.: Neural representation for three-dimensional dose distribution and its applications in precision radiation therapy. Int. J. Radiat. Oncol. Biol. Phys. **114**(3), e552 (2022)
26. Wolterink, J.M., Zwienenberg, J.C., Brune, C.: Implicit neural representations for deformable image registration. In: International Conference on Medical Imaging with Deep Learning, pp. 1349–1359. PMLR (2022)

Investigating Data Memorization in 3D Latent Diffusion Models for Medical Image Synthesis

Salman Ul Hassan Dar[1,2,3(✉)], Arman Ghanaat[1,3], Jannik Kahmann[4], Isabelle Ayx[4], Theano Papavassiliu[2,3,5], Stefan O. Schoenberg[2,4], and Sandy Engelhardt[1,2,3]

[1] Department of Internal Medicine III, Group Artificial Intelligence in CardiovascularMedicine, Heidelberg University Hospital, 69120 Heidelberg, Germany
SalmanUlHassan.Dar@med.uni-heidelberg.de
[2] AI Health Innovation Cluster, Heidelberg, Germany
[3] German Centre for Cardiovascular Research (DZHK), Partner Site Heidelberg/Mannheim, Heidelberg, Germany
[4] Department of Radiology and Nuclear Medicine, University Medical Center Mannheim, Heidelberg University, Theodor-Kutzer-Ufer 1-3, 68167 Mannheim, Germany
[5] First Department of Medicine-Cardiology, University Medical Centre Mannheim, Theodor-Kutzer-Ufer 1-3, 68167 Mannheim, Germany

Abstract. Generative latent diffusion models have been established as state-of-the-art in data generation. One promising application is generation of realistic synthetic medical imaging data for open data sharing without compromising patient privacy. Despite the promise, the capacity of such models to memorize sensitive patient training data and synthesize samples showing high resemblance to training data samples is relatively unexplored. Here, we assess the memorization capacity of 3D latent diffusion models on photon-counting coronary computed tomography angiography and knee magnetic resonance imaging datasets. To detect potential memorization of training samples, we utilize self-supervised models based on contrastive learning. Our results suggest that such latent diffusion models indeed memorize training data, and there is a dire need for devising strategies to mitigate memorization.

Keywords: Deep generative models · Latent diffusion · Data memorization · Patient privacy · Contrastive learning

1 Introduction

Contemporary developments in deep generative modeling have lead to performance leaps in a broad range of medical imaging applications [6,8,10,17–19]. One promising application is generation of novel synthetic images [4,5,9,11,12]. Synthetic images can be used for data diversification by synthesizing samples belonging to underrepresented classes for training of data-driven models or sharing of

A. Mukhopadhyay et al. (Eds.): DGM4MICCAI 2023, LNCS 14533, pp. 56–65, 2024.
https://doi.org/10.1007/978-3-031-53767-7_6

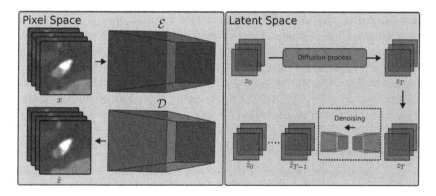

Fig. 1. 3D Latent diffusion models first project 3D sub-volumes onto a lower dimensional latent space for computational efficiency using an encoder. Diffusion models are then trained to gradually denoise the noisy latent space representation. Upon complete denoising, the representation is projected back onto the pixel space using a decoder.

synthetic data for open science without compromising patient privacy. State-of-the art generative models are based on latent diffusion [13]. These models first project data onto a compressed latent space, learn latent space distribution through a gradual denoising process, and synthesize novel latent space samples followed by projection onto a high dimensional pixel space [13]. Despite the ability to synthesize high quality samples, recent studies in computer vision suggest that latent diffusion models (LDMs) are prone to training data memorization [3,15,16]. This can be more critical in medical imaging, where synthesizing real patient data defeats the whole purpose of preserving data privacy. These computer vision studies further suggest that the phenomenon of data memorization is more prevalent in low data regimes [15], which is very often the case in the medical domain. Despite the importance of patient privacy, it is surprising that data memorization in generative models has received little attention in the medical imaging community.

Here, we investigate the memorization capacity of 3D-LDMs in medical images. To this end, we train LDMs to generate 3D volumes (Fig. 1) and compare novel generated samples with real training samples via self-supervised models (Fig. 2) for potential memorization. For assessment, we perform experiments on an in-house photon-counting coronary computed angiography (PCCTA) dataset and a public knee MRI (MRNet) dataset [2]. Our results suggest that LDMs indeed suffer from training data memorization.

1.1 Data Generation via LDMs

LDMs belong to a family of generative models that learn to generate novel realistic samples by denoising normally distributed noise in a compressed lower dimensional latent space [13]. LDMs consist of two models:

Latent Encoding Model. First, an encoder learns to project samples onto lower dimensional latent space. This lower dimensional latent space is typically learned using an autoencoder. The autoencoder is trained to encode the image $x \in \mathbb{R}^{L \times H \times W}$ to a latent space $z \in \mathbb{R}^{L' \times H' \times W'}$ using an encoder $\mathcal{E}$ having parameters $\theta_{\mathcal{E}}$ ($z = \mathcal{E}(x)$), followed by reconstruction via a decoder $\mathcal{D}$ having parameters $\theta_{\mathcal{D}}$ ($\hat{x} = \mathcal{D}(z)$). Overall, the training is performed to minimize the following reconstruction loss function:

$$\mathcal{L}_{rec}(\theta_{\mathcal{E}}, \theta_{\mathcal{D}}) = \mathbb{E}_{p(x)} \left[\|x - \hat{x}\|_1 \right] \tag{1}$$

where $\mathbb{E}_{p(x)}$ denotes expectation with respect to data distribution $p(x)$. Since the encoder and decoder are trained simultaneously with the aim to recover the original image from a lower dimensional representation, the encoder learns to project the data onto semantically meaningful latent space without loosing much information.

Diffusion Model. Afterwards, a deep diffusion probablistic model (DDPM) is trained to recover meaningful latent space from normally distributed noise. DDPMs consist of a forward and reverse diffusion step. In the forward step normally distributed noise is added to the latent representation (z) in small increments. At any time t, the relation between z_t and z_{t-1} can be expressed as:

$$q\left(z_t | z_{t-1}\right) = \mathcal{N}\left(x_t; \sqrt{1 - \beta_t} z_{t-1}, \beta_t \mathbf{I}\right) \tag{2}$$

where β_t is the variance schedule [7] and $q\left(z_t | z_{t-1}\right)$ is the conditional distribution. In the reverse step, a model is trained to approximate $q\left(z_{t-1} | z_t\right)$ as $p_\theta\left(z_{t-1} | z_t\right)$. Once trained, the model can be used to synthesise novel representations (z_0) given $z_T \sim \mathcal{N}\left(0, \mathbf{I}\right)$. The latent code z_0 can then be fed as input to the decoder ($\mathcal{D}$) to generate novel samples from data distribution $p(x)$.

1.2   Memorization Assessment

Although LDMs have outperformed their counterpart generative models in medical image synthesis in terms of image quality and diversity [9,12], their capacity to memorize training samples remains relatively unexplored. This is surprising, considering that one of the main goals of sharing synthetic data is to preserve patient privacy. Memorization of patient data defeats this purpose, and the quality of the synthesized samples becomes secondary.

Since the primary focus of this work is memorization, it is important to first define what constitutes "memorization". Akbar et al. [1] define memorization as a phenomenon where generative models can generate copies of training data samples. However, they do not explicitly define what a "copy" means, and their use of "copy" seems to be limited to a synthesised sample that is identically oriented to a training sample and shares the same anatomical structures with minor differences such as image quality. They detect potential copy candidates

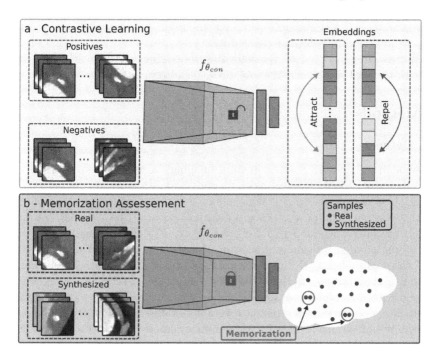

Fig. 2. a - A self-supervised model is trained based on contrastive learning to learn a lower dimensional embedding where augmented versions of the same sample are attracted and different samples are repelled. **b** - The trained model is then used to identify if the synthetic samples are copies of the real samples

by computing pairwise normalized correlation between synthesized and training samples. This fails to take into account that synthetic samples can also be flipped or rotated versions of the training samples, which can easily result as a consequence of data augmentation strategies typically used for training deep models. In our work, we expand the definition of "copy" to further include rotated and flipped versions of a training sample. This definition can further be broadened to include other forms of variations such as slight deformation. However, for simplicity here we limit the additional variations to flipping and rotation. To detect potential copies, we first train a self-supervised model based on the contrastive learning approach (Fig. 2). The aim is to have a low dimensional latent representation of each sample such that the augmented versions of a sample have similar latent representations, and different samples have distinct representations. The model is trained to minimize the following loss function [14]:

$$\mathcal{L}_{con}(\theta_{con}) = \mathbb{E}_{p(x)} \left[\max(0, \left\| f_{\theta_{con}}(x) - f_{\theta_{con}}(x^+) \right\|_2 - \left\| f_{\theta_{con}}(x) - f_{\theta_{con}}(x^-) \right\|_2) \right] \tag{3}$$

where x corresponds to a training volume, x^+ is the similar sample which is an augmented version of x, x^- denotes the dissimilar sample which is just a different volume, and $f_{\theta_{con}}(.)$ is the networks with trainable parameters θ_{con} that maps input x to a low dimensional representation.

After training $f_{\theta_{con}}(.)$, we compare the embeddings of the synthesized samples with the real samples in the low dimensional representational space.

2   Methods

2.1   Datasets

To demonstrate memorization in medical imaging, we selected two datasets covering a range of properties in terms of imaging modalities, organs, resolutions, 3D volume sizes, and dataset sizes. We conducted experiments on in house photon-counting coronary computed tomography angiography (PCCTA) dataset and a publicly available knee MRI (MRNet) dataset [2]. PCCTA images were acquired from 65 patients on a Siemens Naeotom Alpha scanner at the University Medical Centre Mannheim. Ethics approval was approved by the ethics committee of Ethikkommision II, Heidelberg University (ID 2021–659). Images were acquired with a resolution of approximately 0.39 mm × 0.39 mm × 0.42 mm. In all patients, coronary artery plaques were annotated by an expert radiologist. Sub-volumes of size 64 × 64 × 64 surrounding plaques were extracted, resulting in 242 sub-volumes for training and 58 sub-volumes for validation in total. In MRNet, T2-weighted knee MR images of 1130 subjects were analyzed, where 904 subjects were used for training and 226 for validation. All volumes were cropped or zero-padded to have sizes of 256 × 256 × 32. In both datasets, each volume was normalized to have voxel intensity in the range [–1, 1].

2.2   Networks

LDM architecture, training procedures and loss functions were directly adopted from Khader et al. [9]. For the training of the diffusion and autoencoder models, all hyperparameters were matched with the ones selected in Khader et al. [9]. The only exception was the batch size in the diffusion models, which was set to 10 to fit models into the GPU VRAM. For contrastive learning, network architecture was adopted from the encoder in the latent encoding model. The encoder was used to reduce the sub-volume dimensions to 4×4×4 and 8 channels. Afterwards, flattening was performed followed by two densely connected layers to reduce the latent space embeddings to dimensions 32×1. All hyperparameters except for the learning rate and epochs were identical to the latent encoding model. Learning rate and epochs were tuned using a held out validation data.

3   Results

3.1   Memorization Assessment

First, 1000 synthetic PCCTA samples (approx. 4 × training data) were generated using the trained LDM. All synthetic and training samples were then

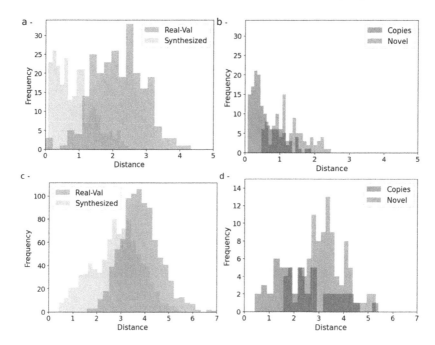

Fig. 3. MSD distributions of copy candidates (Synthesized) and closest validation samples (Real-Val) in **a** - PCCTA and **c** - MRNet datasets are shown. Higher density near zero implies more similarity. MSD distributions of copy candidates in **b** - PCCTA and **d** - MRNet datasets manually annotated as copies or novel samples are shown.

passed through the self-supervised models (Sect. 1.2) to obtain corresponding lower dimensional embeddings. Next, mean square distance (MSD) was computed between all training and synthetic embeddings. For each training sample, the closest synthetic sample was considered as a copy candidate. Figure 3a shows MSD distribution of the candidate copies. To get a better idea of the MSD scale, for each training sample the closest real validation sample in the embedding space was also considered (Fig. 3a). Low values on the x-axis denote lower distance or high similarity. The MSD distribution of synthetic samples is more concentrated near zero compared to the MSD distribution of real validation samples.

To further assess if the candidate synthesized samples are indeed copies, each candidate copy was also labelled manually as a copy or a novel synthetic sample via visual assessment by consensus of two users. As shown in Fig. 3b, most of the candidates with low MSD values are copies. Upon comparing copies with novel samples in 3b, we also observe that 59% of the training data has been memorized. This number is alarming, as it indicates memorization at a large scale. It is also important to note that this percentage is based on just 1000 synthesized samples. Increasing the synthetic samples could lead to an increased number of copies. Figure 4a shows some copy candidates. It can be seen that synthetic samples show stark resemblance with the training samples.

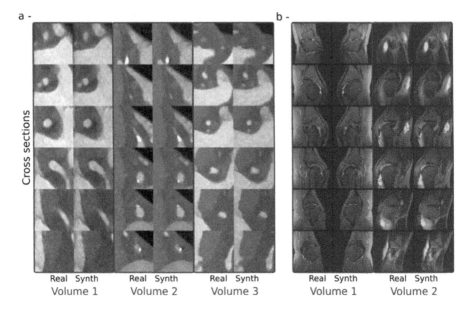

Fig. 4. Representative samples and copy candidates from the **a** - PCCTA and **b** - MRNet dataset are shown. Columns corresponds to real or synthesized (Synth) samples, and rows correspond to six cross sections selected from the sub-volumes. Synthesized samples have a stark resemblance with the real training samples. A copy candidate that is a flipped version of a training sample is also shown (b - Volume 1).

We then assess memorization in the MRNet dataset, which is relatively a larger dataset containing 904 training volumes. 3600 synthetic samples (approx. 4 × training data) were generated using the LDM trained on the MRNet dataset. Figure 5c shows MSD distribution of synthetic candidate copies and validation samples. We observe similar patterns in the MRNet dataset. However, MSD distribution of synthetic candidate copies in MRNet is further away from zero compared to the PCCTA dataset. This can be explained by the training data size, as models with large training datasets get to learn distribution from many diverse samples and thus are less likely to memorize the data. We also annotated 150 randomly selected copy candidates as copy or novel samples (Fig. 5d). We find 33% of the copy candidates to be copies. Figure 4-b also shows representative samples.

3.2   Data Augmentation

We also analyze the effect of data augmentation on memorization by comparing MSD distribution of copy candidates generated by models trained with augmentation (augmented models) and without augmentation (non-augmented models) on the PCCTA dataset. Figure 5 compares the MSD distributions. MSD distribution of the non-augmented model tends to have higher density near zero.

Overall, 41% of the training dataset is memorized in augmented models compared to 59% in non-augmented models. This suggests that the non-augmented models tend to memorize more than the augmented models. One possible explanation could be artificial expansion of datasets through augmentation, which can in turn lead to fewer repetitions of identical forms of a sample during training.

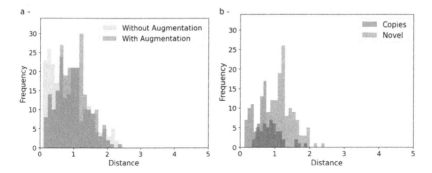

Fig. 5. a - MDS distribution of copy candidates for models trained with and without data augmentation. **b** - MSD distributions of copy candidates manually annotated as copies or novel samples on models trained with data augmentation.

4   Discussion

There has been a considerable amount of focus on generative models in medical image synthesis. Here, we tried to assess if these models actually learn to synthesize novel samples as opposed to memorizing training samples. Our results suggest that LDMs indeed memorize samples from the training data. This can have broad implications in the medical imaging community, since leakage of patient data in the form of medical images can lead to violation of patient privacy.

An interesting future prospect could be to understand the underlying reasons leading to memorization. This will also enable us in suggesting potential solutions to mitigate memorization. Somepalli et al. [16] suggests that data duplication during training could be an important factor, as a repeated sample is seen many more times during training. They further suggest that unconditional models primarily suffer from data memorization in low data regimes, which is similar to what we observe when we compare memorization in PCCTA and MRNet datasets. Nonetheless, it is an important research direction which warrants future work. Another interesting future direction could be to compare memorization in different diffusion models such as 2D diffusion models, text conditioned diffusion models, and 3D diffusion models with different training settings.

To our knowledge, this is the first study assessing memorization in 3D LDMs for medical image synthesis. Another independent study recently investigated memorization in deep diffusion models [1]. There are several differences between our study and Akbar et al. : 1) Akbar et al. trained 2D models on medical images, whereas we trained 3D models which are more coherent with the nature of the medical images. 2) Akbar et al. is based on diffusion models in pixel space, which are not easily applicable to 3D medical images due to high computational demands. To the contrary, here we used LDMs, which first project the data onto a low dimension latent space to reduce computational complexity while ensuring that the relevant semantic information is preserved. 3) Akbar et al. used correlation between images in the pixel space to assess memorization. While this approach detects identical copies, it cannot detect augmented or slightly different copies. Here, we trained a self-supervised model that can also account for augmented versions of the training samples, which might be missed by computing regular correlations in pixel space. 4) We also assessed memorization on models that were trained using augmented data. This is a more realistic scenario while training deep models on medical images due to data scarcity.

Acknowledgment. This work was supported through state funds approved by the State Parliament of Baden-Württemberg for the Innovation Campus Health + Life Science Alliance Heidelberg Mannheim, BMBF-SWAG Project 01KD2215D, and Informatics for life project through Klaus Tschira Foundation. The authors also gratefully acknowledge the data storage service SDS@hd supported by the Ministry of Science, Research and the Arts Baden-Württemberg (MWK) and the German Research Foundation (DFG) through grant INST 35/1314-1 FUGG and INST 35/1503-1 FUGG. The authors also acknowledge support by the state of Baden-Württemberg through bwHPC and the German Research Foundation (DFG) through grant INST 35/1597-1 FUGG.

References

1. Akbar, M.U., Wang, W., Eklund, A.: Beware of diffusion models for synthesizing medical images - a comparison with GANs in terms of memorizing brain tumor images (2023)
2. Bien, N., et al.: Deep-learning-assisted diagnosis for knee magnetic resonance imaging: development and retrospective validation of MRNet. PLOS Med. **15**(11), 1–19 (2018). https://doi.org/10.1371/journal.pmed.1002699
3. Carlini, N., et al.: Extracting training data from diffusion models (2023)
4. Dorjsembe, Z., Odonchimed, S., Xiao, F.: Three-dimensional medical image synthesis with denoising diffusion probabilistic models. In: Medical Imaging with Deep Learning (2022)
5. Engelhardt, S., Sharan, L., Karck, M., Simone, R.D., Wolf, I.: Cross-domain conditional generative adversarial networks for stereoscopic hyperrealism in surgical training. In: Shen, D., et al. (eds.) MICCAI 2019. LNCS, vol. 11768, pp. 155–163. Springer, Cham (2019). https://doi.org/10.1007/978-3-030-32254-0_18
6. Güngör, A., et al.: Adaptive diffusion priors for accelerated MRI reconstruction. Med. Image Anal. 102872 (2023). https://doi.org/10.1016/j.media.2023.102872
7. Ho, J., Jain, A., Abbeel, P.: Denoising diffusion probabilistic models. In: Larochelle, H., Ranzato, M., Hadsell, R., Balcan, M.F., Lin, H. (eds.) Advances in Neural Information Processing Systems, vol. 33, pp. 6840–6851. Curran Associates, Inc. (2020)
8. Kazerouni, A., Aghdam, E.K., Heidari, M., Azad, R., Fayyaz, M., Hacihaliloglu, I., Merhof, D.: Diffusion models in medical imaging: a comprehensive survey. Med. Image Anal. **88**, 102846 (2023). https://doi.org/10.1016/j.media.2023.102846

9. Khader, F., et al.: Denoising diffusion probabilistic models for 3D medical image generation. Sci. Rep. **13**(1), 7303 (2023). https://doi.org/10.1038/s41598-023-34341-2

10. Özbey, M., Dalmaz, O., Dar, S.U., Bedel, H.A., Özturk, c., Güngör, A., Çukur, T.: Unsupervised medical image translation with adversarial diffusion models. IEEE Trans. Med. Imaging 1 (2023). https://doi.org/10.1109/TMI.2023.3290149

11. Pfeiffer, M., et al.: Generating large labeled data sets for laparoscopic image processing tasks using unpaired image-to-image translation. In: Shen, D., et al. (eds.) MICCAI 2019. LNCS, vol. 11768, pp. 119–127. Springer, Cham (2019). https://doi.org/10.1007/978-3-030-32254-0_14

12. Pinaya, W.H.L., et al.: Brain imaging generation with latent diffusion models. In: Mukhopadhyay, A., Oksuz, I., Engelhardt, S., Zhu, D., Yuan, Y. (eds.) Deep Generative Models, DGM4MICCAI 2022, LNCS, vol. 13609, pp 117–126. Springer, Cham (2022). https://doi.org/10.1007/978-3-031-18576-2_12

13. Rombach, R., Blattmann, A., Lorenz, D., Esser, P., Ommer, B.: High-resolution image synthesis with latent diffusion models. In: 2022 IEEE/CVF Conference on Computer Vision and Pattern Recognition (CVPR), pp. 10674–10685 (2022). https://doi.org/10.1109/CVPR52688.2022.01042

14. Schroff, F., Kalenichenko, D., Philbin, J.: FaceNet: a unified embedding for face recognition and clustering. In: Proceedings of the IEEE Conference on Computer Vision and Pattern Recognition (CVPR), June 2015

15. Somepalli, G., Singla, V., Goldblum, M., Geiping, J., Goldstein, T.: Diffusion art or digital forgery? investigating data replication in diffusion models. In: Proceedings of the IEEE/CVF Conference on Computer Vision and Pattern Recognition (CVPR), pp. 6048–6058, June 2023

16. Somepalli, G., Singla, V., Goldblum, M., Geiping, J., Goldstein, T.: Understanding and mitigating copying in diffusion models (2023)

17. Wolleb, J., Bieder, F., Sandkühler, R., Cattin, P.C.: Diffusion models for medical anomaly detection. In: Wang, L., Dou, Q., Fletcher, P.T., Speidel, S., Li, S. (eds.) Medical Image Computing and Computer Assisted Intervention - MICCAI 2022, pp. 35–45. Springer Nature Switzerland, Cham (2022). https://doi.org/10.1007/978-3-031-16452-1_4

18. Wolleb, J., Sandkühler, R., Bieder, F., Valmaggia, P., Cattin, P.C.: Diffusion models for implicit image segmentation ensembles. In: Konukoglu, E., Menze, B., Venkataraman, A., Baumgartner, C., Dou, Q., Albarqouni, S. (eds.) Proceedings of The 5th International Conference on Medical Imaging with Deep Learning. Proceedings of Machine Learning Research, vol. 172, pp. 1336–1348. PMLR, 06–08 July 2022

19. Yi, X., Walia, E., Babyn, P.: Generative adversarial network in medical imaging: a review. Med. Image Anal. **58**, 101552 (2019). https://doi.org/10.1016/j.media.2019.101552

ViT-DAE: Transformer-Driven Diffusion Autoencoder for Histopathology Image Analysis

Xuan Xu[✉], Saarthak Kapse, Rajarsi Gupta, and Prateek Prasanna

Stony Brook University, New York, USA
xuaxu@cs.stonybrook.edu,
{saarthak.kapse,prateek.prasanna}@stonybrook.edu,
Rajarsi.Gupta@stonybrookmedicine.edu

Abstract. Generative AI has received substantial attention in recent years due to its ability to synthesize data that closely resembles the original data source. While Generative Adversarial Networks (GANs) have provided innovative approaches for histopathological image analysis, they suffer from limitations such as mode collapse and overfitting in discriminator. Recently, Denoising Diffusion models have demonstrated promising results in computer vision. These models exhibit superior stability during training, better distribution coverage, and produce high-quality diverse images. Additionally, they display a high degree of resilience to noise and perturbations, making them well-suited for use in digital pathology, where images commonly contain artifacts and exhibit significant variations in staining. In this paper, we present a novel approach, namely ViT-DAE, which integrates vision transformers (ViT) and diffusion autoencoders for high-quality histopathology image synthesis. This marks the first time that ViT has been introduced to diffusion autoencoders in computational pathology, allowing the model to better capture the complex and intricate details of histopathology images. We demonstrate the effectiveness of ViT-DAE on three publicly available datasets. Our approach outperforms recent GAN-based and vanilla DAE methods in generating realistic images.

Keywords: Histopathology · Diffusion Autoencoders · Vision Transformers

1 Introduction

Over the last few years, generative models have sparked significant interest in digital pathology [9]. The objective of generative modeling techniques is to create synthetic data that closely resembles the original or desired data distribution. The synthesized data can improve the performance of various downstream

Supplementary Information The online version contains supplementary material available at https://doi.org/10.1007/978-3-031-53767-7_7.

tasks [24], eliminating the need for obtaining large-scale costly expert annotated data. Methods built on Generative Adversarial Networks (GANs) have provided novel approaches to address various challenging histopathological image analysis problems including stain normalization [20], artifact removal [3], representation learning [1], data augmentation [27,29], etc. However, GAN models experience mode collapse and limited latent size [17]. Their discriminator is prone to overfitting while producing samples from datasets with imbalanced classes [28]. This results in lower quality image synthesis.

Diffusion models, on the other hand, are capable of producing images that are more diverse. They are also less prone to overfitting compared to GANs [28]. Recently, Denoising Diffusion Probabilistic models (DDPMs) [8] and score-based generative models [23] have shown promise in computer vision. Ho et al. present high quality image synthesis using diffusion probabilistic models [8]. Latent diffusion models have been proposed for high-resolution image synthesis which are widely applied in super resolution, image inpainting, and semantic scene synthesis [18]. These models have been shown to achieve better synthetic image quality compared to GANs [4]. Denoising Diffusion Implicit models (DDIMs) construct a class of non-Markovian diffusion processes which makes sampling from reverse process much faster [22]. This modification in the forward process preserves the goal of DDPM and allows for deterministically encoding an image to the noise map. Unlike DDPMs [15], DDIMs enable control over image synthesis owing to the latent space flexibility (attribute manipulation) [17].

Motivation: Diffusion models have the potential to make a significant impact in computational pathology. Compared with other generative models, diffusion models are generally more stable during training. Unlike natural images, histopathology images are intricate and harbor rich contextual information which can make GANs difficult to train and suffer from issues like mode collapse where the generator produces limited and repetitive outputs failing to capture the complex diversity in histopathology image distribution. Diffusion models, on the contrary, are more stable and provide better distribution coverage leading to high fidelity diverse images. They are considerably more resistant to perturbations and noise, which is crucial in digital pathology because images routinely contain artifacts and exhibit large variations in staining [10]. With desirable attributions in high quality image generation, diffusion models have the capacity to enhance various endeavors in computational pathology including image classification, data augmentation, and super resolution. Despite their potential benefits, diffusion models remain largely unexplored in computational pathology.

Recently, Preechakul et al. [17] proposed a diffusion autoencoder (DAE) framework which encodes natural images into a representation using semantic encoder and uses the resulting semantic subcode as the condition in the DDIM image decoder. The encoding of histopathology images, however, presents a significant challenge, primarily because these images contain intricate microenvironments of tissues and cells, which have complex and diverse spatial arrangements. Thus, it is imperative to use a semantic encoder that has a high capacity to represent and understand the complex and global spatial structures

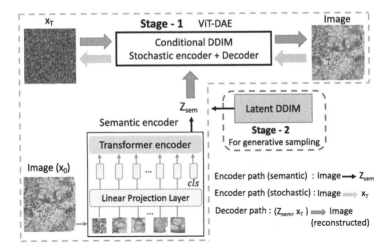

Fig. 1. Overview of the proposed ViT-DAE framework. **Training:** In Stage-1, an input image is encoded into a semantic representation by the ViT. This representation is taken as the condition for the conditional DDIM to decode the noisy image. In Stage-2, a latent DDIM is trained to learn the distribution of semantic representations of data. **Generative sampling:** We synthesize the semantic representations from the latent DDIM and feed it to the conditional DDIM along with randomly initialized noisy image to generate new histopathology samples.

present in histopathology images. Towards this direction, we propose to introduce vision transformer (ViT) [5] as the semantic encoder instead of their convolutional neural network (CNN) counterpart in DAE; our proposed method is called ViT-DAE. The self-attention [25] mechanism in ViT allows to better capture global contextual information through long-range interactions between the regions/patches of images. A previous study [16] has shown that 1) ViTs perform better than CNNs and are comparable to humans on shape recognition, 2) ViTs are more robust against perturbations and image corruptions. Based on these findings, ViT presents a promising solution in encoding meaningful and rich representations of complex and noisy tissue structures, where shape and spatial arrangement of biological entities form two crucial motifs. Hence we hypothesize that a transformer-based semantic encoder in DAE would result in higher quality histopathology image synthesis.

To summarize our main contributions, **(1)** We are the first to introduce conditional DDIM in histopathology image analysis, and **(2)** We enhance the conditional DDIM by incorporating ViT as a semantic encoder, enabling it to holistically encode complex phenotypic layout specific to histopathology. We demonstrate the effectiveness of ViT-DAE on three public datasets; it outperforms recent GAN-based and vanilla DAE methods, in generating better images.

2 Proposed Method

To generate histopathology images that are meaningful and of diagnostic quality, we propose ViT-DAE, a framework that utilizes a transformer-enhanced diffusion autoencoder. Our method consists of two stages of **training**. *Stage-1* comprises i) a ViT-based semantic encoder, which captures the global semantic information of an *input image*, and ii) a conditional DDIM which is an autoencoder, takes in input the semantic representation by ViT as a condition and the noisy image to reconstruct the *input image*. In *Stage-2* with the frozen semantic encoder, a latent DDIM is trained to learn the distribution of semantic representation of the data in the latent space. Following this, for **Generative sampling**, first the latent DDIM is fed with a noisy vector, outputting a synthesized sample from the learned semantic representation distribution. This, along with a randomly initialized noisy image is then fed to the conditional DDIM for image generation. An overview of the proposed framework is shown in Fig. 1.

Vision-Transformer Enhanced Semantic Encoder. An input image $\mathbf{x}_0$ is encoded into a semantic representation $\mathbf{z}_{sem}$ via our ViT based semantic encoder $\mathbf{z}_{sem} = \mathrm{Enc}_\phi(\mathbf{x}_0)$. We split the input image into patches and apply a linear projection followed by interacting them in the transformer encoder. The output class token, cls, is projected to dimension $d = 512$ via a linear layer, and then is used as a condition for the decoder part of DAE. The semantic representations from the input images encoded by the ViT provide an information-rich latent space which is then utilized as the condition in DDIM following [17].

Conditional DDIM. A Gaussian diffusion process can be described as the gradual addition of small amount of Gaussian noise to input images in T steps which leads to a sequence of noisy images $\mathbf{x}_1, ..., \mathbf{x}_T$ [8,17]. At a given time t (out of T), the diffusion process can be defined as $q(\mathbf{x}_t|\mathbf{x}_{t-1}) = \mathcal{N}(\sqrt{1-\beta_t}\mathbf{x}_{t-1}, \beta_t\mathbf{I})$, where β_t are the noise level hyperparameters. The corresponding noisy image of $\mathbf{x}_0$ at time t is another Gaussian $q(\mathbf{x}_t|\mathbf{x}_0) = \mathcal{N}(\sqrt{\alpha_t}\mathbf{x}_0, (1-\alpha_t)\mathbf{I})$ where $\alpha_t = \prod_{s=1}^t (1-\beta_s)$. This is followed by learning of the generative reverse process, i.e., the distribution $p(\mathbf{x}_{t-1}|\mathbf{x}_t)$ [8,17]. DDIM [22] proposes this reverse process as a deterministic generative process, given by:

$$\mathbf{x}_{t-1} = \sqrt{\alpha_{t-1}}\left(\frac{\mathbf{x}_t - \sqrt{1-\alpha_t}\epsilon_\theta^t(\mathbf{x}_t)}{\sqrt{\alpha_t}}\right) + \sqrt{1-\alpha_{t-1}}\epsilon_\theta^t(\mathbf{x}_t) \tag{1}$$

where $\epsilon_\theta^t(\mathbf{x}_t)$, proposed by [8], is a function which takes the noisy image $\mathbf{x}_t$ and predicts the noise using a UNet [19]. The inference distribution is given by:

$$q(\mathbf{x}_{t-1}|\mathbf{x}_t, \mathbf{x}_0) = \mathcal{N}\left(\sqrt{\alpha_{t-1}}\mathbf{x}_0 + \sqrt{1-\alpha_{t-1}}\frac{\mathbf{x}_t - \sqrt{\alpha_t}\mathbf{x}_0}{\sqrt{1-\alpha_t}}, \mathbf{0}\right) \tag{2}$$

The conditional DDIM decoder takes an input in the form of $\mathbf{z} = (\mathbf{z}_{sem}, \mathbf{x}_T)$ to generate the output images. Conditional DDIM leverages the reverse process defined in Eqs. 3, 4 to model $p_\theta(\mathbf{x}_{t-1}|\mathbf{x}_t, \mathbf{z}_{\text{sem}})$ to match the inference distribution $q(\mathbf{x}_{t-1}|\mathbf{x}_t, \mathbf{x}_0)$ defined in Eqs. 2.

$$p_\theta(\mathbf{x}_{0:T}|\mathbf{z}_{\text{sem}}) = p(\mathbf{x}_T) \prod_{t=1}^{T} p_\theta(\mathbf{x}_{t-1}|\mathbf{x}_t, \mathbf{z}_{\text{sem}}) \tag{3}$$

$$p_\theta(\mathbf{x}_{t-1}|\mathbf{x}_t, \mathbf{z}_{\text{sem}}) = \begin{cases} \mathcal{N}(\mathbf{f}_\theta(\mathbf{x}_1, 1, \mathbf{z}_{\text{sem}}), \mathbf{0}) & \text{if } t = 1 \\ q(\mathbf{x}_{t-1}|\mathbf{x}_t, \mathbf{f}_\theta(\mathbf{x}_t, t, \mathbf{z}_{\text{sem}})) & \text{otherwise} \end{cases} \tag{4}$$

where $\mathbf{f}_\theta$ in Eqs. 4 is parameterized as the noise prediction network $\epsilon_\theta(\mathbf{x}_t, t, \mathbf{z}_{\text{sem}})$ from Song et al. [22]:

$$\mathbf{f}_\theta(\mathbf{x}_t, t, \mathbf{z}_{\text{sem}}) = \frac{1}{\sqrt{\alpha_t}}(\mathbf{x}_t - \sqrt{1 - \alpha_t}\epsilon_\theta(\mathbf{x}_t, t, \mathbf{z}_{sem})) \tag{5}$$

This network is a modified version of a UNet from [4].

Generative Sampling. To generate images from the diffusion autoencoder, a latent DDIM is leveraged to learn the semantic representation distribution of $\mathbf{z}_{\text{sem}} = \text{Enc}_\phi(\mathbf{x}_0)$, $\mathbf{x}_0 \sim p(\mathbf{x}_0)$. We follow the framework from [17] to leverage the deep MLPs (10–20 layers) with skip connections as the latent DDIM network. Loss $\mathcal{L}_{\text{latent}}$ is optimized during training with respect to latent DDIM's parameter, ω:

$$\mathcal{L}_{\text{latent}} = \sum_{t=1}^{T} \mathbb{E}_{\mathbf{z}_{\text{sem}}, \epsilon_t}\left[||\epsilon_\omega(\mathbf{z}_{\text{sem},t}, t) - \epsilon_t||_1\right] \tag{6}$$

where $\epsilon_t \in \mathbb{R}^d \sim \mathcal{N}(\mathbf{0}, \mathbf{I})$, $\mathbf{z}_{\text{sem},t} = \sqrt{\alpha_t}\mathbf{z}_{\text{sem}} + \sqrt{1 - \alpha_t}\epsilon_t$ and T is the same as our conditional image decoder.

The semantic representations are normalized to zero mean and unit variance before being fed to the latent DDIM, to model the semantic representation distribution. Generative sampling using diffusion autoencoders involves three steps. First, we sample the $\mathbf{z}_{\text{sem}}$ from the latent DDIM which learns the distribution of semantic representations and unnormalizes it. Then we sample $\mathbf{x}_T \sim \mathcal{N}(\mathbf{0}, \mathbf{I})$. Finally we decode $\mathbf{z} = (\mathbf{z}_{\text{sem}}, \mathbf{x}_T)$ via the conditional DDIM image decoder. Note that for generating class-specific images, an independent latent DDIM model is trained on semantic distribution for each class. Whereas for class-agnostic sampling, just one latent DDIM is trained on the complete cohort.

3   Experiments and Results

3.1   Datasets and Implementation Details

In this study, we utilized 4 datasets. The self-supervised (SSL) pretraining of the semantic encoder - vision transformer (ViT) [5] using DINO [2] framework is carried out on TCGA-CRC-DX [12], NCT-CRC-HE-100K [11], and PCam [26]. The corresponding pre-trained ViT models are then used as semantic encoders for training separate diffusion autoencoders [17] on Chaoyang [30], NCT-CRC-HE-100K [11], and PCam [26] datasets, respectively. Since the Chaoyang dataset contains only a few thousand images, DINO pre-training is conducted on another dataset (TCGA-CRC-DX) consisting of images from the same organ (colon). For NCT-CRC and PCam, the official train split provided for each dataset is used for self-supervision and diffusion autoencoder training. **TCGA-CRC-DX** [12] consists of images of tumor tissue of colorectal cancer (CRC) WSIs in the TCGA database (N=368505). All the images are of size 512×512 pixels (px). This dataset is just utilized for self-supervised pre-training. **Chaoyang** [30] contains a total of 6160 colon cancer images of size 512×512 px. These images are assigned one of the four classes - normal, serrated, adenocarcinoma, and adenoma by the consensus of three pathologists. The official train split is used to train the diffusion autoencoder as well as patch classification model; the official test split [30] is used to report the downstream classification performance. **NCT-CRC-HE-100K** [11] contains 100k (224×224 px) non-overlapping images from histological images of human colorectal cancer (CRC) and normal (H&E)-stained tissue. There are nine classes in this dataset. **PCam** [26] includes 327,680 images (96×96 px) taken from histopathologic scans of lymph node sections in breast. Each image has a binary annotation indicating the presence of metastatic tissue. **Environment:** Our framework is built in PyTorch 1.8.1 and trained on two Quadro RTX 8000 GPUs. We use ViT-Small as our transformer encoder. It is pretrained via DINO using default parameters [2]. All images are resized to 224×224 px for SSL. In contrast, due to memory constraints for generative model training, the images are scaled to 128×128 px. To optimize the diffusion autoencoder, we adopted default parameters and configurations from DAE [17].

Evaluation Metrics: We employ Frechet inception distance (**FID**), Improved Precision, and Improved Recall to evaluate the similarity between the distribution of synthesized images and real images [7,14]. For FID computation [21], the real and synthesized images are fed into an Inception V3 model [13] to extract features from pool_3 layer. The FID method then calculates the difference between mean and standard deviation from these features. A lower FID score indicates a higher similarity between the distributions. For NCT-CRC and PCam, we computed FID scores between all the real images from training set and our 50k generated images. For Chaoyang, FID is computed between all the training images and the generated 3k images. Improved Precision (**IP**) and Recall (**IR**) estimate the distribution of real images and synthesized images by forming explicit, non-parameteric representations of the manifolds [14]. IP describes the probability that a random generated image falls within the support of the real image manifold. Conversely, IR is defined as the probability that a random real image belongs to the generated image manifold.

For the downstream classification analysis with generated samples by ViT-DAE on Chaoyang dataset, we report accuracy as well as class-wise F1 score.

Manifold Visualization: Motivated by [14], we generate manifolds to demonstrate the superior performance of our method. We employ Principal Component Analysis (PCA) to reduce the dimensionality of the representation space; the top two PCs represent the transformed 2D feature space. We compute the radii for each feature vector by fitting the manifold algorithm to the transformed space. We then generate a manifold by plotting circles with the 2D vector as their centers and their corresponding radii as radius. A lower FID score indicates better IP, meaning the generated images are visually closer to the real images. It also suggests better IR, indicating that the generated images cover a larger portion of the real image manifold and exhibit higher diversity. This provides a comprehensive visualization and allows for an intuitive understanding of the learned representations.

Table 1. Comparison of FID, IP, IR on three datasets.

Dataset	NCT-CRC			PCam			Chaoyang		
Metric	FID ↓	IP ↑	IR ↑	FID ↓	IP ↑	IR ↑	FID ↓	IP ↑	IR ↑
VQ-GAN [6]	27.86	0.57	0.26	15.99	0.56	0.22	51.35	0.43	**0.53**
DAE [17]	14.91	0.58	0.30	39.42	0.32	0.28	**35.69**	0.50	0.44
ViT-DAE (ours)	**12.14**	**0.60**	**0.40**	**13.39**	**0.60**	**0.44**	36.18	**0.51**	0.50

3.2 Results

We compared two contemporary methods, Diffusion Autoencoder (DAE) [17] and VQ-GAN [6], which are built on the latest advances in diffusion and GAN-based algorithms, respectively. Table 1 compares the quality of synthesized images produced by ViT-DAE with other methods.

Quantitative Analysis: Images produced by ViT-DAE have the lowest FID scores and highest IP and IR for the NCT-CRC and PCam datasets; the results are comparable with the other two approaches, particularly DAE, on the Chaoyang dataset. This consistent improvement may be attributed to ViT's replacement of the convolution-based semantic encoder. ViT has a far greater capacity to learn the contextual information and long-range relationships in the images than its CNN counterparts. As a result, a ViT-based semantic encoder could more effectively capture high level semantic information and provide more meaningful representation as the condition of DDIM, which improves the quality of generated images. CNN-based conditional DDIM faces the difficulty of encoding the complex spatial layouts in histopathology images. Since the CNN-based semantic encoder and the ViT-based encoder consist of comparable number of parameters (24.6M and 21.6M, respectively), we attribute this improvement to the superior global modeling of ViTs.

Qualitative Analysis: The similarity between distribution coverage of real and generated images is visually evaluated using manifold visualization in Fig. 2.

For the NCT-CRC and PCam datasets, our synthesized images have a closer distribution to the real image manifold compared to VQ-GAN and DAE. For Chaoyang (in supplementary), DAE and ViT-DAE generate distributions that are very comparable to real images while outperforming the generated manifold from VQ-GAN. We also provide class-wise samples along with pathologist's interpretations of synthesized images generated by ViT-DAE in Fig. 3. We can see that our method captures the distribution sufficiently well and generates reasonably plausible class-specific images (more in supplementary). Here we summarize our pathologist's impressions on *"How phenotypically real are the synthesized images?"* for the different classes in NCT-CRC. *Normal mucosa:* Realistic in terms of cells configured as glands as the main structural element of colonic mucosa, location of nuclei at the base of glands with apical mucin adjacent to the central lumen. *Lymphocytes:* Realistic in terms of cell contours, sizes, color, texture, and distribution. *Mucus:* Color and texture realistic for mucin (mucoid material secreted from glands). *Smooth Muscle:* Realistic with central nuclei in elongated spindle cells with elastic collagen fibers and no striations. *Cancer associated stroma:* Realistic in terms of reactive stroma with inflammatory infiltrate (scattered lymphs, neutrophil) and increased cellularity of stromal cells. *Tumor:* Realistic with disordered growth via nuclear crowding with a diversity of larger than normal epithelial (glandular epithelial) cells and irregular shaped cells.

Downstream Analysis: To assess the efficacy of synthesized images, we conducted a proof-of-concept investigation by designing a classification task on

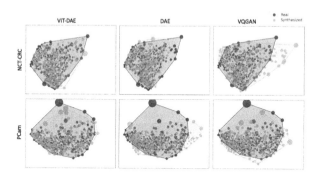

Fig. 2. Manifold visualization. Higher overlap indicates greater similarity.

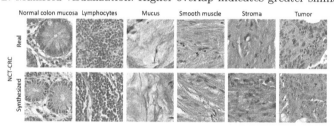

Fig. 3. Real and synthesized images using ViT-DAE on NCT-CRC

the Chaoyang dataset. We trained a classification model exclusively on class-conditioned synthesized images and evaluated its performance on real images from official test split [30]. The performance was comparable to that of a classification model trained solely on real images from provided train split. Our results demonstrate that a hybrid training approach, where we combine real and synthesized images, can significantly improve the classifier performance on small-scale datasets like Chaoyang, especially for underrepresented minority classes by up to 4–5%. Our findings (results in supplementary) highlight the potential utility of synthesized images in improving the performance of downstream tasks.

4   Conclusion

Our study proposes a novel approach that combines conditional DDIM with a vision transformer-based semantic encoder (ViT-DAE) for high-quality histopathology image synthesis. Our method outperforms recent GAN-based and vanilla DAE methods, demonstrating its effectiveness in generating more realistic and diverse histopathology images. The use of vision transformers in semantic encoder of DAE enables the holistic encoding of complex layouts specific to histo-pathology images, making it a promising solution for future research on image synthesis in digital pathology.

References

1. Boyd, J., Liashuha, M., Deutsch, E., Paragios, N., Christodoulidis, S., Vakalopoulou, M.: Self-supervised representation learning using visual field expansion on digital pathology. In: Proceedings of the IEEE/CVF International Conference on Computer Vision (2021)
2. Caron, M., et al.: Emerging properties in self-supervised vision transformers. In: Proceedings of the IEEE/CVF International Conference on Computer Vision, pp. 9650–9660 (2021)
3. Dahan, C., Christodoulidis, S., Vakalopoulou, M., Boyd, J.: Artifact removal in histopathology images. arXiv preprint arXiv:2211.16161 (2022)
4. Dhariwal, P., Nichol, A.: Diffusion models beat GANs on image synthesis. Adv. Neural. Inf. Process. Syst. **34**, 8780–8794 (2021)
5. Dosovitskiy, A., et al.: An image is worth 16x16 words: transformers for image recognition at scale. arXiv preprint arXiv:2010.11929 (2020)
6. Esser, P., Rombach, R., Ommer, B.: Taming transformers for high-resolution image synthesis. In: Proceedings of the IEEE/CVF Conference on Computer Vision and Pattern Recognition, pp. 12873–12883 (2021)
7. Heusel, M., Ramsauer, H., Unterthiner, T., Nessler, B., Hochreiter, S.: GANs trained by a two time-scale update rule converge to a local nash equilibrium. In: Advances in Neural Information Processing Systems, vol. 30 (2017)
8. Ho, J., Jain, A., Abbeel, P.: Denoising diffusion probabilistic models. Adv. Neural. Inf. Process. Syst. **33**, 6840–6851 (2020)
9. Jose, L., Liu, S., Russo, C., Nadort, A., Di Ieva, A.: Generative adversarial networks in digital pathology and histopathological image processing: a review. J. Pathol. Inf. **12**(1), 43 (2021)

10. Kanwal, N., Pérez-Bueno, F., Schmidt, A., Engan, K., Molina, R.: The devil is in the details: whole slide image acquisition and processing for artifacts detection, color variation, and data augmentation: a review. IEEE Access **10**, 58821–58844 (2022)
11. Kather, J.N., Halama, N., Marx, A.: 100,000 histological images of human colorectal cancer and healthy tissue. Zenodo10 (2018)
12. Kather, J.N., et al.: Deep learning can predict microsatellite instability directly from histology in gastrointestinal cancer. Nat. Med. **25**(7), 1054–1056 (2019)
13. Kynkäänniemi, T., Karras, T., Aittala, M., Aila, T., Lehtinen, J.: The role of imagenet classes in fr\'echet inception distance. arXiv preprint arXiv:2203.06026 (2022)
14. Kynkäänniemi, T., Karras, T., Laine, S., Lehtinen, J., Aila, T.: Improved precision and recall metric for assessing generative models. In: Advances in Neural Information Processing Systems, vol. 32 (2019)
15. Moghadam, P.A., et al.: A morphology focused diffusion probabilistic model for synthesis of histopathology images. In: Proceedings of the IEEE/CVF Winter Conference on Applications of Computer Vision, pp. 2000–2009 (2023)
16. Naseer, M.M., Ranasinghe, K., Khan, S.H., Hayat, M., Shahbaz Khan, F., Yang, M.H.: Intriguing properties of vision transformers. Adv. Neural. Inf. Process. Syst. **34**, 23296–23308 (2021)
17. Preechakul, K., Chatthee, N., Wizadwongsa, S., Suwajanakorn, S.: Diffusion autoencoders: toward a meaningful and decodable representation. In: Proceedings of the IEEE/CVF Conference on Computer Vision and Pattern Recognition, pp. 10619–10629 (2022)
18. Rombach, R., Blattmann, A., Lorenz, D., Esser, P., Ommer, B.: High-resolution image synthesis with latent diffusion models. In: Proceedings of the IEEE/CVF Conference on Computer Vision and Pattern Recognition, pp. 10684–10695 (2022)
19. Ronneberger, O., Fischer, P., Brox, T.: U-net: convolutional networks for biomedical image segmentation. In: Navab, N., Hornegger, J., Wells, W.M., Frangi, A.F. (eds.) MICCAI 2015, Part III. LNCS, vol. 9351, pp. 234–241. Springer, Cham (2015). https://doi.org/10.1007/978-3-319-24574-4_28
20. Runz, M., Rusche, D., Schmidt, S., Weihrauch, M.R., Hesser, J., Weis, C.A.: Normalization of he-stained histological images using cycle consistent generative adversarial networks. Diagn. Pathol. **16**(1), 1–10 (2021)
21. Seitzer, M.: pytorch-fid: FID Score for PyTorch. https://github.com/mseitzer/pytorch-fid (August 2020), version 0.3.0
22. Song, J., Meng, C., Ermon, S.: Denoising diffusion implicit models. arXiv preprint arXiv:2010.02502 (2020)
23. Song, Y., Sohl-Dickstein, J., Kingma, D.P., Kumar, A., Ermon, S., Poole, B.: Score-based generative modeling through stochastic differential equations. arXiv preprint arXiv:2011.13456 (2020)
24. Tellez, D., et al.: Quantifying the effects of data augmentation and stain color normalization in convolutional neural networks for computational pathology. Med. Image Anal. **58**, 101544 (2019)
25. Vaswani, A., et al.: Attention is all you need. In: Advances in Neural Information Processing Systems, vol. 30 (2017)
26. Veeling, B.S., Linmans, J., Winkens, J., Cohen, T., Welling, M.: Rotation equivariant CNNs for digital pathology. In: Frangi, A.F., Schnabel, J.A., Davatzikos, C., Alberola-López, C., Fichtinger, G. (eds.) MICCAI 2018. LNCS, vol. 11071, pp. 210–218. Springer, Cham (2018). https://doi.org/10.1007/978-3-030-00934-2_24

27. Wei, J., et al.: Generative image translation for data augmentation in colorectal histopathology images. Proc. Mach. Learn. Res. **116**, 10 (2019)

28. Xiao, Z., Kreis, K., Vahdat, A.: Tackling the generative learning trilemma with denoising diffusion GANs. arXiv preprint arXiv:2112.07804 (2021)

29. Xue, Y., et al.: Selective synthetic augmentation with histogan for improved histopathology image classification. Med. Image Anal. **67**, 101816 (2021)

30. Zhu, C., Chen, W., Peng, T., Wang, Y., Jin, M.: Hard sample aware noise robust learning for histopathology image classification. IEEE Trans. Med. Imaging **41**(4), 881–894 (2021)

Anomaly Guided Generalizable Super-Resolution of Chest X-Ray Images Using Multi-level Information Rendering

Vamshi Vardhan Yadagiri$^{(\boxtimes)}$, Sekhar Reddy, and Angshuman Paul

Indian Institute of Technology Jodhpur, Jheepasani, India
{vardhan.1,reddy.11,apaul}@iitj.ac.in

Abstract. Single image super-resolution (SISR) methods aim to generate a high-resolution image from the corresponding low-resolution images. Such methods may be useful in improving the resolution of medical images including chest x-rays. Medical images with superior resolution may subsequently lead to an improved diagnosis. However, SISR methods for medical images are relatively rare. We propose a SISR method for chest x-ray images. Our method uses multi-level information rendering by utilizing the cue about the abnormality present in the images. Experiments on publicly available datasets show the superiority of the proposed method over several state-of-the-art approaches.

Keywords: Super-resolution · Multi-level information rendering · Chest x-ray · Anomaly guided

1 Introduction

A superior resolution of x-ray images may lead to an improved diagnosis compared to their low-resolution counterparts. Chest X-ray images are one of the most widely used imaging techniques in medical diagnosis, and enhancing the resolution of these images can result in improved and faster diagnoses. Single image super-resolution (SISR) techniques [1], have proven to be effective in improving the resolution of medical images, potentially leading to better diagnoses.

However, the use of SISR methods for medical images including chest x-rays is relatively unexplored. Among the few existing methods, in [1], the authors embed a modified squeeze and excitation block in EDSR [2] for the super-resolution of retinal images. Zhang *et al.* have proposed a method for the super-resolution of medical images from different modalities [3]. A progressive GAN architecture has been used for the super-resolution of pathology and magnetic resonance images [4]. See [5] for the use of multi-level skip connections to perform medical image super-resolution. The less abundance of SISR methods for medical images may be primarily attributed to the ill-posed nature of the SISR problem [6]. In this paper, we propose an anomaly guided SISR method for chest x-rays that uses Multi-label information rendering. We make the following major contributions:

© The Author(s), under exclusive license to Springer Nature Switzerland AG 2024
A. Mukhopadhyay et al. (Eds.): DGM4MICCAI 2023, LNCS 14533, pp. 77–85, 2024.
https://doi.org/10.1007/978-3-031-53767-7_8

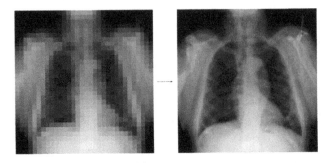

Fig. 1. 32 times super-resolution of a chest X-ray image from 32 × 32 pixels to 1024 × 1024 pixels using the proposed model

- We propose a method for SISR of chest X-ray images using Multi-level Information Rendering (MLIR).
- We propose a novel loss function that utilizes the abnormality if present in the chest x-rays for SISR.
- We design a two-stage training process where initially (stage 1) we train utilizing only MLIR integrated into the baseline model. Subsequently, in stage 2 of training, we exploit the information about the abnormality present in images.

The rest of the paper is organized as follows. In Sect. 2, we discus our method followed by experiments and results in. Section 3. Finally, we conclude the paper in Sect. 4.

2   Method

We use a Generative Adversarial Network (GAN) [7] to implement multi-label information rendering for SISR. Our model consists of a generator and a discriminator. Our generator tries to construct a super-resolved (SR) image from its low-resolution (LR) counterpart. The discriminator tries to differentiate the generated SR images from ground truth high-resolution (HR) images.

2.1   The Generator

Our generator is built on the generator architecture of ESRGAN [8] which in turn was built upon the SRGAN [9]. A block diagram of the generator is presented in Fig. 2. The generator utilizes the concept of generative adversarial networks along with the ResNet [10] architecture. We introduce multi-Level information rendering (MLIR) to design our generator.

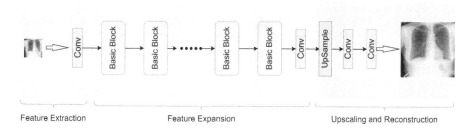

Fig. 2. A block diagram of the ESRGAN [8] generator architecture (Conv: convolution layer with kernel size 3 × 3, Upsample: upsampling layer with pixel-shuffle layers that increases the dimensions by 4 times).

Multi-level Information Rendering Blocks: Our MLIR approach is inspired by [11]. It is based on the intuition that different levels of abstraction of the input data may contain complementary information. Combining this information may help in better performance. We implement MLIR blocks to improve the performance of the generator by capturing more information from LR images. Our goal is to retain more low-level features for the super-resolution of the LR images. We also expect this to result in giving us improved finer-level details.

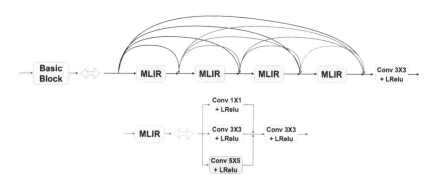

Fig. 3. The structure and contents of the Basic Block and MLIR block. Conv Y × Y represents a convolution layer with filter size Y × Y and LReLu represents the Leaky Rectified Linear Activation function.

Each MLIR block consists of three parallel convolutional blocks. By connecting the convolutional blocks in parallel rather than in series, we can capture various levels of information from each block and effectively combine them to produce high-quality feature maps. We connect a series of MLIR blocks to construct a basic block as shown in Fig. 3. Our generator is similar to that of Fig. 2 with the difference that our basic blocks are designed using MLIR blocks, unlike the ESRGAN.

2.2   The Discriminator

We use the Relativistic Discriminator from the Relativistic average GAN (RaGAN) [12]. This discriminator is used in ESRGAN. The Relativistic Discriminator computes a probability score that takes into account the relationship between the generator's output and the ground truth. Specifically, instead of just comparing the generated sample and the real sample, the Relativistic Discriminator compares the difference between the generated sample and the average of the real samples, to the difference between a real sample and the average of the generated samples.

2.3   The Loss Function

If there is any abnormality present in the LR images, we want to enhance the details of the abnormality in the super-resolved images. To that end, we introduce a bounding box loss ($\mathcal{L}_{BB}$). Bounding box loss is the generator loss of [8] computed using the region containing the anomaly in the HR image and that region in the corresponding generated SR image. We also compute the loss for the whole image ($\mathcal{L}_I$) by computing the generator loss. Our generator loss is similar to that of ESRGAN [8]. This loss is computed using the HR images and the corresponding SR images. Therefore, the total loss of the generator is

$$\mathcal{L}_T = \mathcal{L}_I + \lambda \mathcal{L}_{BB} \tag{1}$$

where $\lambda \in (0, 1)$ is a hyperparameter indicating the relative weight of the aforementioned losses. For the discriminator, we use a loss similar to that of ESRGAN. Thus, the information about the anomaly present in the x-ray image help in the training of our model. Based on the validation performance on 50 images, a value of 0.2 is chosen for λ.

2.4   Training

We use images from the NIH Chest X-ray dataset [13] for training our model. From the original images, we create LR images through bicubic downsampling and normalization of the original images. The HR images were created by the normalization of the original images. We use a two-stage training procedure as follows.

Stage 1: During the first stage of training, we minimize loss $\mathcal{L}_I$ for the generator. After the LR and HR image pairs were generated, the size of all the LR images are 256 × 256 whereas the HR images retained their original dimensions of 1024 × 1024. These images were then given as input to the main model and the model was run for 200 epochs on PyTorch version 22.02 using an Nvidia DGX-2 server that had 16 V100 GPUs and ran on Ubuntu 18.04 LTS. From these 200 epochs, the best model was taken using validation by 50 images.

Stage 2. After the best model was obtained, images from the NIH dataset that had bounding boxes were used to train that model for another 100 epochs. In this training, we minimize the loss function of (1) for the generator. Once the model is trained, we use it for generating super-resolved x-ray images.

3 Experiments and Results

3.1 Datasets

We perform experiments on several publicly available chest x-ray datasets. The primary dataset is the NIH Chest X-ray dataset [13]. This dataset consists of a total of 112120 images. Out of these, there are more than 800 images that have bounding boxes showing the regions containing abnormalities. We also use two other datasets, namely CheXpert [14] dataset containing 224316 images, and VinBigData Chest X-ray [15] dataset containing 18,000 images. All of the images in the VinBigData Chest X-ray dataset have bounding boxes.

We use 30000 images from the NIH dataset for training, 50 images for validation, and 25595 images for testing. In order to look into the generalizability of the training, the model trained on the NIH dataset is tested on the CheXpert and VinBigData datasets. We have used all the frontal images from the CheXpert dataset (191027 images) and all the images of the VinBigData dataset for testing. For CheXpert dataset, we have performed only ×4 super-resolution.

3.2 Comparative Analysis of Performance

We evaluate the performance of our proposed model in terms of peak signal-to-noise ratio (PSNR) (the higher the better) and mean squared error (MSE) (the lower the better). We compare the performance of our model with several state-of-the-art models, namely, ESRGAN [8], EDSR [2], and USRNET [16]. The comparative performances of different methods for ×4 super-resolution are presented in Table 1. Notice that our method achieves superior results compared to most of the approaches. We can also observe that our model trained on the NIH dataset shows consistent performance across other datasets. This shows the generalizability of the proposed model.

We also perform experiments for ×32 super-resolution. The comparative performances are presented in Table 2. Notice that our method outperforms ESRGAN for this task as well. Visual results of ×32 super-resolution using our model are presented in Fig. 4 and Fig. 5. Figure 6 presents a visual comparison of the generated images from different models.

Table 1. Performances of different methods trained on NIH dataset for ×4 super-resolution (LR input: 256 × 256 pixels, SR output: 1024 × 1024 pixels) on various test datasets).

		ESRGAN [8]	USRNET [16]	EDSR [2]	Proposed
NIH [13]	PSNR(dB)	39.61	40.01	40.31	40.72
	MSE	7.66	6.49	6.05	5.84
CheXpert [14]	PSNR(dB)	29.52	28.04	32.86	32.17
	MSE	72.53	102.11	33.66	40.2
VinBig [15]	PSNR(dB)	38.29	38.79	40.57	39.98
	MSE	9.64	8.59	5.70	6.53

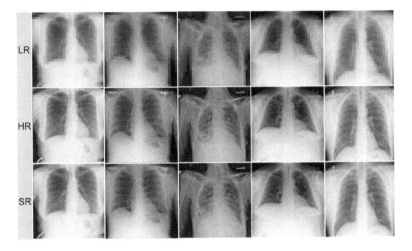

Fig. 4. Visual results of ×32 super-resolution using the proposed method on the NIH chest X-ray dataset (LR: low-resolution image, SR: super-resolved images by our model, HR: ground truth high-resolution image).

3.3 Ablation Studies

Impact of the Bounding Box Loss: We first look into the impact of the bounding box loss. To that end, we perform experiments excluding the bounding box loss in our model (abbreviated as 'With MLIR'). The result of this experiment is presented in Table 2. Notice that for both ×4 and ×32 super-resolution, the proposed model outperforms 'With MLIR '. This shows the impact of bounding box loss in the proposed model. Visual results of this experiment are presented in Fig. 7.

Impact of the MLIR Blocks: Next, we evaluate the impact of the MLIR blocks in our model. For this purpose, we perform experiments without the MLIR blocks (abbreviated as 'With BB Loss ') on the NIH dataset. The results of this

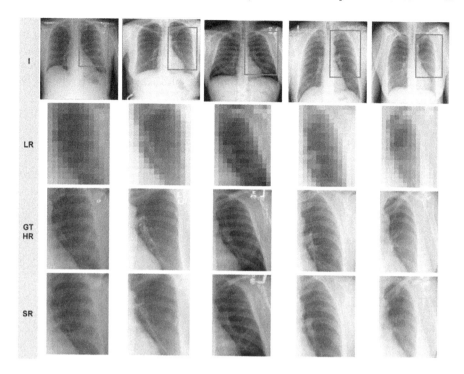

Fig. 5. Visual results obtained on the NIH chest X-ray dataset using the proposed model. This figure shows the detailed results in small selected regions of the images (I: input image, LR: low-resolution image, GT HR: the high-resolution image of the selected region of ground truth image, SR: super-resolved image.)

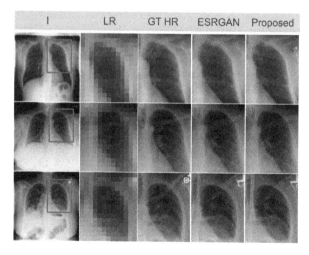

Fig. 6. Visual comparisons showing the output of different models for ×32 super-resolution in small selected regions of the images (I: input image, LR: low-resolution image, GT HR: the high-resolution image of the selected region of ground truth image).

Table 2. PSNR and MSE values for different models. ESRGAN is the baseline model. With BB loss represents our model trained with bounding box loss but without MLIR blocks., With MLIR represents our model without bounding box loss. The results are presented for ×4 (LR input: 256 × 256 pixels, SR output: 1024 × 1024 pixels) and ×32 (LR input: 32 × 32 pixels, SR output: 1024 × 1024 pixels) super-resolution for the NIH test dataset.

		ESRGAN [8]	With BB loss	With MLIR	Proposed
×4	PSNR(dB)	39.61	40.28	40.31	40.72
	MSE	7.66	6.54	6.52	5.84
×32	PSNR(dB)	32.32	32.75	32.70	32.84
	MSE	39.80	35.92	36.57	35.26

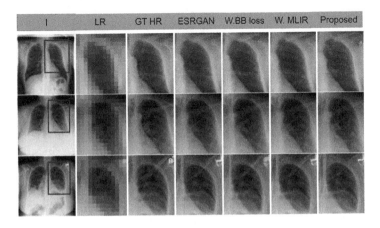

Fig. 7. Visual comparisons showing the impact of various components of the proposed method in small selected regions of the images (I: input image, LR: low-resolution, GT HR: the high-resolution image of the selected region of ground truth image, ESR-GAN: baseline ESRGAN model without any added components, W.BB loss: model with Bounding box loss, W.MLIR: model with MLIR, Proposed: our model with both Bounding box loss and MLIR integrated).

experiment are reported in Table 2. Notice that our method outperforms this ablation study. This shows the importance of the MLIR blocks. Visual results of this experiment are presented in Fig. 7.

4 Conclusions

We propose a SISR method for chest x-ray images. Our method is designed using multi-level information rendering. We exploit the information about the region containing anomaly to design the loss function. The results on publicly available datasets show the generalizability of the proposed method. In the future, we will evaluate the quality of the super-resolved images on various downstream tasks.

Additionally, we will extend our method to other radiology imaging modalities, such as CT. We will also look into 3D information fusion in this context.

Acknowledgement. The authors thank the National Institutes of Health Clinical Center for providing the NIH dataset.

References

1. Bing, X., Zhang, W., Zheng, L., Zhang, Y.: Medical image super resolution using improved generative adversarial networks. IEEE Access **7**, 145030–145038 (2019)
2. Lim, B., Son, S., Kim, H., Nah, S., Mu Lee, K.: Enhanced deep residual networks for single image super-resolution. In: Proceedings of the IEEE Conference on Computer Vision and Pattern Recognition Workshops, pp. 136–144 (2017)
3. Zhang, S., Liang, G., Pan, S., Zheng, L.: A fast medical image super resolution method based on deep learning network. IEEE Access **7**, 12319–12327 (2018)
4. Mahapatra, D., Bozorgtabar, B., Garnavi, R.: Image super-resolution using progressive generative adversarial networks for medical image analysis. Comput. Med. Imaging Graph. **71**, 30–39 (2019)
5. Qiu, D., Zheng, L., Zhu, J., Huang, D.: Multiple improved residual networks for medical image super-resolution. Futur. Gener. Comput. Syst. **116**, 200–208 (2021)
6. Yang, W., Zhang, X., Tian, Y., Wang, W., Xue, J.-H., Liao, Q.: Deep learning for single image super-resolution: a brief review. IEEE Trans. Multimedia **21**(12), 3106–3121 (2019)
7. Goodfellow, I., et al.: Generative adversarial networks. Commun. ACM **63**(11), 139–44 (2020)
8. Wang, X., et al.: Esrgan: enhanced super-resolution generative adversarial networks. In: Proceedings of the European Conference on Computer Vision (ECCV) Workshops (2018)
9. Ledig, C., et al.: Photo-realistic single image super-resolution using a generative adversarial network. In: Proceedings of the IEEE Conference on Computer Vision and Pattern Recognition, pp. 4681–4690 (2017)
10. He, K., Zhang, X., Ren, S., Sun, J.: Deep residual learning for image recognition. In: Proceedings of the IEEE Conference on Computer Vision and Pattern Recognition, pp. 770–778 (2016)
11. Ding, X., Zhang, X., Ma, N., Han, J., Ding, G., Sun, J.: Repvgg: making VGG-style convnets great again. In: Proceedings of the IEEE/CVF Conference on Computer Vision and Pattern Recognition, pp. 13733–13742 (2021)
12. Jolicoeur-Martineau, A.: The relativistic discriminator: a key element missing from standard GAN. arXiv preprint arXiv:1807.00734. 2 July 2018
13. Wang, X., Peng, Y., Lu, L., Lu, Z., Bagheri, M., Summers, R.: Hospital-scale chest x-ray database and benchmarks on weakly-supervised classification and localization of common thorax diseases. In: IEEE CVPR, vol. 7, p. 46 (2017)
14. Irvin, J., et al.: Chexpert: a large chest radiograph dataset with uncertainty labels and expert comparison. In: Proceedings of the AAAI Conference on Artificial Intelligence, vol. 33, no. 01, pp. 590–597, 17 July 2019
15. Nguyen, H.Q., et al.: VinDr-CXR: an open dataset of chest x-rays with radiologist's annotations. Scientific Data. **9**(1), 429 (2022)
16. Zhang, K., Gool, L.V., Timofte, R.: Deep unfolding network for image super-resolution. In: Proceedings of the IEEE/CVF Conference on Computer Vision and Pattern Recognition, pp. 3217–3226 (2020)

Importance of Aligning Training Strategy with Evaluation for Diffusion Models in 3D Multiclass Segmentation

Yunguan Fu[1,2]([✉])(iD), Yiwen Li[3](iD), Shaheer U. Saeed[1](iD),
Matthew J. Clarkson[1](iD), and Yipeng Hu[1,3](iD)

[1] University College London, London, UK
yunguan.fu.18@ucl.ac.uk
[2] InstaDeep, London, UK
[3] University of Oxford, Oxford, UK

Abstract. Recently, denoising diffusion probabilistic models (DDPM) have been applied to image segmentation by generating segmentation masks conditioned on images, while the applications were mainly limited to 2D networks without exploiting potential benefits from the 3D formulation. In this work, we studied the DDPM-based segmentation model for 3D multiclass segmentation on two large multiclass data sets (prostate MR and abdominal CT). We observed that the difference between training and test methods led to inferior performance for existing DDPM methods. To mitigate the inconsistency, we proposed a recycling method which generated corrupted masks based on the model's prediction at a previous time step instead of using ground truth. The proposed method achieved statistically significantly improved performance compared to existing DDPMs, independent of a number of other techniques for reducing train-test discrepancy, including performing mask prediction, using Dice loss, and reducing the number of diffusion time steps during training. The performance of diffusion models was also competitive and visually similar to non-diffusion-based U-net, within the same compute budget. The JAX-based diffusion framework has been released at https://github.com/mathpluscode/ImgX-DiffSeg.

Keywords: Image Segmentation · Diffusion Model · Prostate MR · Abdominal CT

1 Introduction

Multiclass segmentation is one of the most basic tasks in medical imaging, one that arguably benefited the most from deep learning. Although different model architectures [15,25] and training strategies [6,16] have been proposed for specific clinical applications, U-net [21] trained through supervised training remains

---

Supplementary Information The online version contains supplementary material available at https://doi.org/10.1007/978-3-031-53767-7_9.

the state-of-the-art and an important baseline for many [9]. Recently, denoising diffusion probabilistic models (DDPM) have been demonstrated to be effective in a variety of image synthesis tasks [7], which can be further guided by a scoring model to generate conditioned images [5] or additional inputs [8]. These generative modelling results are followed by image segmentation, where the model generates segmentation masks by progressive denoising from random noise. During training, DDPM is provided with an image and a noise-corrupted segmentation mask, generated by a linear interpolation between the ground-truth and a sampled noise. The model is then tasked to predict the sampled noise [2,13,26,27] or the ground-truth mask [4].

However, existing applications have been mainly based on 2D networks and, for 3D volumetric medical images, slices are segmented before obtaining the assembled 3D segmentation. Challenges are often encountered for 3D images. First, the diffusion model requires image and noise-corrupted masks as input, leading to an increased memory footprint resulting in limited batch size and potentially excessive training time. For instance, the transformer-based architecture becomes infeasible without reducing model size or image, given clinically or academically accessible hardware with limited memory. Second, most diffusion models assume a denoising process of hundreds of time steps for training and inference, the latter of which in particular leads to prohibitive inference time (e.g., days on TPUs/GPUs).

This work addresses these issues by aligning training with evaluation processes via recycling. As discussed in multiple studies [4,13,14,28], noise does not necessarily disrupt the shape of ground truth masks and morphological features may be preserved in noise-corrupted masks during training. By training with recycling (Fig. 1), the prediction from the previous steps is used as input, i.e. rather than the ground truth used in existing methods, for noisy mask sampling. This proposed training process emulates the test process since the input is also from the previous predictions at inference time for diffusion models, without access to ground truth. Furthermore, this work directly predicts ground-truth masks instead of sampled noise [27]. This facilitates the direct use of Dice loss in addition to cross-entropy during training, as opposed tp L_2 loss on noise. Lastly, instead of denoising with at least hundreds of steps as in most existing work, we propose a five-step denoising process for both training and inference, resorting to resampling variance scheduling [17].

With extensive experiments in two of the largest public multiclass segmentation applications, prostate MR (589 images) and abdominal CT images (300 images) [9,16], we demonstrated a statistically significant improvement (between 0.015 and 0.117 in Dice score) compared to existing DDPMs. Compared to non-diffusion supervised learning, diffusion models reached a competitive performance (between 0.008 and 0.015 in Dice), with the same computational cost. With high transparency and reproducibility, avoiding selective results under different conditions, we conclude that the proposed recycling strategy using mask prediction setting with Dice loss should be the default configuration for 3D segmentation applications with diffusion models. We release the first unit-tested JAX-based diffusion segmentation framework at https://github.com/mathpluscode/ImgX-DiffSeg.

2   Related Work

The diffusion probabilistic model was first proposed by Sohl-Dickstein et al. [23] as a generative model for image sampling with a forward noising process. Ho et al. [7] proposed a reverse denoising process that estimates the sampled error, achieving state-of-the-art performance in unconditioned image synthesis at the time. Different conditioning methods were later proposed to guide the sampling process toward a desired image class or prompt text, using gradients from an external scoring model [5,19]. Alternatively, Ho et al. [8] showed that guided sampling can be achieved by providing conditions during training. Diffusion models have been successfully applied in medical imaging applications to synthesise images of different modalities, such as unconditioned lung X-Ray and CT [1], patient-conditioned brain MR [18], temporal cardiac MR [11], and pathology/sequence-conditioned prostate MR [22]. The synthesised images have been shown to benefit pre-training self-supervised models [10,22] or support semi-supervised learning [28].

Besides image synthesis, Baranchuk et al. [3] used pre-trained diffusion models' intermediate feature maps to train pixel classifiers for segmentation, showing these unsupervised models capture semantics that can be extended for image segmentation especially when training data is limited. Alternatively, Amit et al. [2] performed progressive denoising from random sampled noise to generate segmentation masks instead of images for microscopic images [2]. At each step, the model takes a noise-corrupted mask and an image as input and predicts the sampled noise. Similar approaches have been also applied to thyroid lesion segmentation for ultrasound images [27] and brain tumour segmentation for MR images with different network architectures [26,27]. Empirically, multiple studies [4,13,28] found the noise-corrupted mask generation, via linear interpolation between ground-truth masks and noise, retained morphological features during training, causing potential data leakage. Chen et al. [4] therefore added noises to mask analog bit and tuned its scaling. Young et al. [28], on the other hand, tuned the variance and scaling of added normal noise to reduce information contained in noised masks. Furthermore, Kolbeinsson et al. [13] proposed recursive denoising instead of directly using ground truth for noise-corrupted mask generation.

These works, although using different methods, are all addressing a similar concern: the diffusion model training process is different from its evaluation process, which potentially hinges the efficient learning. Moreover, most published diffusion-model-based segmentation applications have been based on 2D networks. We believe such discrepancy would be more significant when applying 3D networks to volumetric images due to the increased difficulty, resulting in longer training and larger compute cost. In this work, building on these recent developments, we focus on a consistent train-evaluate algorithm for efficient training of diffusion models in 3D medical image segmentation applications.

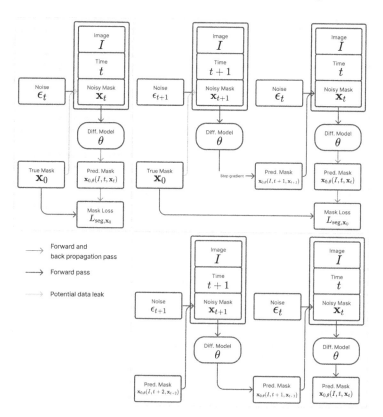

Fig. 1. Illustration of training with and without recycling and inference, using mask prediction. For training without recycling (top left), the noisy mask $\mathbf{x}_t$ is calculated using the ground truth mask. For training with recycling (top right), $\mathbf{x}_t$ is calculated using prediction from the previous step, which is similar to the inference (bottom right).

3   Method

3.1   DDPM for Segmentation

The denoising diffusion probabilistic models (DDPM) [7,12,17,23] consider a *forward* process: given sample $\mathbf{x}_0 \sim q(\mathbf{x}_0)$, noisy $\mathbf{x}_t$ for $t = 1, \cdots, T$ follows a multivariate normal distribution, $q(\mathbf{x}_t \mid \mathbf{x}_{t-1}) = \mathcal{N}(\mathbf{x}_t; \sqrt{1 - \beta_t}\mathbf{x}_{t-1}, \beta_t \mathbf{I})$, where $\beta_t \in [0, 1]$. Given a sufficiently large T, $\mathbf{x}_T$ approximately follows an isotropic multivariate normal distribution $\mathcal{N}(\mathbf{x}_T; \mathbf{0}, \mathbf{I})$. A *reverse* process is then defined to denoise $\mathbf{x}_t$ at each step, for $t = T, \cdots, 1$, $p_\theta(\mathbf{x}_{t-1} \mid \mathbf{x}_t) = \mathcal{N}(\mathbf{x}_{t-1}; \boldsymbol{\mu}_\theta(\mathbf{x}_t, t), \tilde{\beta}_t \mathbf{I})$, with a predicted mean $\boldsymbol{\mu}_\theta(\mathbf{x}_t, t)$ and a variance schedule $\tilde{\beta}_t \mathbf{I}$. $\tilde{\beta}_t = \frac{1 - \bar{\alpha}_{t-1}}{1 - \bar{\alpha}_t}\beta_t$, where $\alpha_t = 1 - \beta_t$, $\bar{\alpha}_t = \prod_{s=0}^{t} \alpha_s$. $\boldsymbol{\mu}_\theta(\mathbf{x}_t, t)$ can be modelled in two different ways,

$$\mu_\theta(\mathbf{x}_t, t) = \frac{\sqrt{\bar{\alpha}_{t-1}}\beta_t}{1 - \bar{\alpha}_t}\mathbf{x}_{0,\theta}(\mathbf{x}_t, t) + \frac{1 - \bar{\alpha}_{t-1}}{1 - \bar{\alpha}_t}\sqrt{\alpha_t}\mathbf{x}_t, \quad \text{(Predict } \mathbf{x}_0) \tag{1}$$

$$\mu_\theta(\mathbf{x}_t, t) = \frac{1}{\sqrt{\alpha_t}}(\mathbf{x}_t - \frac{\beta_t}{\sqrt{1 - \bar{\alpha}_t}}\epsilon_{t,\theta}(\mathbf{x}_t, t)), \quad \text{(Predict noise } \epsilon_t) \tag{2}$$

where $\mathbf{x}_{0,\theta}(\mathbf{x}_t, t)$ or $\epsilon_{t,\theta}(\mathbf{x}_t, t)$ are the learned neural network.

For segmentation, $\mathbf{x}$ represents a transformed probability with values in $[-1, 1]$. Particularly, $\mathbf{x}_0 \in \{1, -1\}$ are binary-valued, transformed from mask. $\mathbf{x}_{0,\theta}$ and $\mathbf{x}_t$ ($t \geq 0$) have values in $[-1, 1]$. Moreover, the networks $\mathbf{x}_{0,\theta}(I, \mathbf{x}_t, t)$ or $\epsilon_{t,\theta}(I, \mathbf{x}_t, t)$ takes one more input I, representing the image to segment.

3.2   Recycling

During training, existing methods samples $\mathbf{x}_t$ by interpolating noise ϵ_t and ground-truth $\mathbf{x}_0$, which results in a certain level of data leak [4,28]. Kolbeinsson et al. [13] proposed recursive denoising, which performed T steps on each image progressively to use model's predictions at previous steps. However, this extends the training length T times. Instead, in this work, for each image, the time step t is randomly sampled and the model's prediction $\mathbf{x}_{0,\theta}$ from the previous time step $t + 1$ is recycled to replace ground-truth (Fig. 1).

Similar reuses of the model's predictions have been previously applied in 2D image segmentation [4], however $\mathbf{x}_{0,\theta}$ was fed into the network along with $\mathbf{x}_t$ which requires additional memories and still has data leak risks. A further difference to these previous approaches is that, rather than stochastic recycling, usually with a probability of 50%, it is always applied throughout the training (which was empirically found to lead to more stable and performant model training). Formally, the recycling technique at a sampled step t is as follows,

$$\mathbf{x}_{t+1} = \sqrt{\bar{\alpha}_{t+1}}\mathbf{x}_0 + \sqrt{1 - \bar{\alpha}_{t+1}}\epsilon_{t+1}, \quad \text{(Noise mask generation for } t + 1) \tag{3}$$

$$\mathbf{x}_{0,\theta} = \text{StopGradient}(\mathbf{x}_{0,\theta}(I, t + 1, \mathbf{x}_{t+1})), \quad \text{(Mask prediction)} \tag{4}$$

$$\mathbf{x}_t = \sqrt{\bar{\alpha}_t}\mathbf{x}_{0,\theta} + \sqrt{1 - \bar{\alpha}_t}\epsilon_t, \quad \text{(Noise mask generation for } t) \tag{5}$$

where $\mathbf{x}_{0,\theta}$ is the predicted segmentation mask from $t+1$ using ground-truth, with gradient stopping. ϵ_t and ϵ_{t+1} are two independently sampled noises. Recycling can be applied to models predicting noise (see supplementary materials Sect. 1 for derivation and illustration).

3.3   Loss

Given noised mask $\mathbf{x}_t$, time t, and image I, the loss can be,

$$L_{\text{seg},\mathbf{x}_0}(\theta) = \mathbb{E}_{t,\mathbf{x}_0,\epsilon_t,I} L_{\text{seg}}(\mathbf{x}_0, \mathbf{x}_{0,\theta}), \tag{6}$$

$$L_{\text{seg},\epsilon_t}(\theta) = \mathbb{E}_{t,\mathbf{x}_0,\epsilon_t,I} \|\epsilon_t - \epsilon_{t,\theta}\|_2^2, \tag{7}$$

where model predict noise $\epsilon_{t,\theta}$ or mask $\mathbf{x}_{0,\theta}$ and $\mathcal{L}_{\text{seg}}$ represents a segmentation-specific loss, such as Dice loss or cross-entropy loss. t is sampled from 1 to T, $\epsilon_t \sim \mathcal{N}(\mathbf{0}, \mathbf{I})$, and $\mathbf{x}_t(\mathbf{x}_0, \epsilon_t) = \sqrt{\bar{\alpha}_t}\mathbf{x}_0 + \sqrt{1 - \bar{\alpha}_t}\epsilon_t$. When model predicts noise, segmentation loss can still be used as the mask can be inferred via interpolation.

3.4 Variance Resampling

During training or inference, given a variance schedule $\{\beta_t\}_{t=1}^{T}$ for $T = 1000$, a subsequence $\{\beta_k\}_{k=1}^{K}$ for $K = 5$ can be sampled with $\{t_k\}_{k=1}^{K}$, where $t_K = T$, $t_1 = 1, \beta_k = 1 - \frac{\bar{\alpha}_{t_k}}{\bar{\alpha}_{t_{k-1}}}, \tilde{\beta}_k = \frac{1 - \bar{\alpha}_{t_{k-1}}}{1 - \bar{\alpha}_{t_k}}\beta_{t_k}$. α_k and $\bar{\alpha}_k$ are recalculated correspondingly.

4 Experiment Setting

Prostate MR. The data set[1] [16] contains 589 T2-weighted image-mask pairs for 8 anatomical structures. The images were randomly split into non-overlapping training, validation, and test sets, with 411, 14, 164 images in each split, respectively. All images were normalised, resampled, and centre-cropped to an image size of $256 \times 256 \times 48$, with a voxel dimension of $0.75 \times 0.75 \times 2.5$ (mm).

Abdominal CT. The data set[2] [9] provides 200 and 100 CT image-mask pairs for 15 abdominal organs in training and validation sets. The validation set was randomly split into non-overlapping validation and test sets, with 10 and 90 images, respectively. HU values were clipped to $[-991, 362]$ for all images. Images were then normalised, resampled and centre-cropped to an image size of $192 \times 128 \times 128$, with a voxel dimension of $1.5 \times 1.5 \times 5.0$ (mm).

Implementation. 3D U-nets have four layers with 32, 64, 128, and 256 channels. For diffusion models, noise-corrupted masks and images were concatenated along feature channels and time was encoded using sinusoidal positional embedding [20]. Random rotation, translation and scaling were adopted for data augmentation during training. The segmentation-specific loss function is by default the sum of cross-entropy and foreground-only Dice loss. When predicting noise, the L_2 loss has a weight of 0.1 [27]. All models were trained with AdamW for 12500 steps and a warmup cosine learning rate schedule. Hyper-parameter were configured empirically without extensive tuning. Binary Dice score and 95% Hausdorff distance in mm (HD), averaged over foreground classes, were reported. Paired Student's t-tests were performed on Dice score to test statistical significance between model performance with $\alpha = 0.01$. Experiments were performed using bfloat16 mixed precision on TPU v3-8. The code is available at https://github.com/mathpluscode/ImgX-DiffSeg.

---

[1] https://zenodo.org/record/7013610#.ZAkaXuzP2rM.
[2] https://zenodo.org/record/7155725#.ZAkbe-zP2rO.

5 Results

Recycling. The performance of the proposed diffusion model is summarised in Table 1. With recycling, the diffusion-based 3D models reached a Dice score of 0.830 and 0.801 which is statistically significantly ($p < 0.001$) higher than baseline diffusion with 0.815 and 0.753, for prostate MR and abdominal CT respectively. Example predictions were provided in Fig. 2.

Diffusion vs Non-diffusion. The diffusion model is also compared with U-net trained via standard supervised learning in Table 2. Within the same computing budget, the diffusion-based 3D model is competitive with its non-diffusion counterpart. The results in Fig. 2 are also visually comparable. The difference, however, remains significant with $p < 0.001$.

Ablation Studies. Comparisons were performed on prostate MR data set for other modifications (see Table 3), including: 1) predicting mask instead of noise 2) using Dice loss in addition to cross-entropy, 3) using five steps denoising process during training. Improvements in Dice score between 0.09 and 0.117 were observed for all modifications (all p-values < 0.001). The largest improvement was observed when the model predicted segmentation masks instead of noise. The results were found consistent with the consistency model [24], which requests diffusion models' predictions of $\mathbf{x}_0$ from different time steps to be similar. Such requirement is implicitly met in our applications as the segmentation loss requires the prediction to be consistent with the ground truth mask given an image. As a result, the predictions from different time steps shall be consistent.

Limitation. In general, all methods tend to perform better for large regions of interest (ROI), and there is a significant correlation (Spearman $r > 0.8$ and $p <$

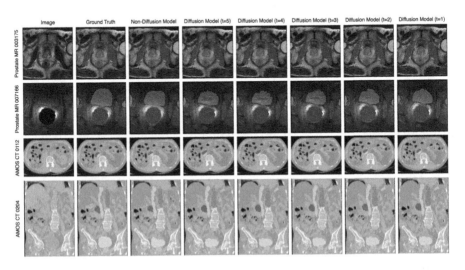

Fig. 2. Example predictions from diffusion and non-diffusion U-net models.

0.01) between ROI (regions of interest) area and mean Dice score per ROI/class, indicating room of future improvement by addressing small ROIs.

Table 1. DDPM with recycling.

Recycling	Prostate MR		Abdominal CT	
	Dice Score	Hausdorff Dist.	Dice Score	Hausdorff Dist.
N	0.815 ± 0.095	5.485 ± 1.069	0.753 ± 0.131	9.526 ± 2.232
Y	0.830 ± 0.094	5.424 ± 1.176	0.801 ± 0.109	9.125 ± 2.564

Table 2. Comparison between diffusion models and non-diffusion models.

Diffusion	Prostate MR		Abdominal CT	
	Dice score	Hausdorff Dist.	Dice score	Hausdorff Dist.
N	0.838 ± 0.088	5.197 ± 1.184	0.816 ± 0.100	9.091 ± 2.475
Y	0.830 ± 0.094	5.424 ± 1.176	0.801 ± 0.109	9.125 ± 2.564

Table 3. Ablation results (prostate MR). HD stands for Hausdorff distance.

(a) Mask or noise prediction

Output	Dice Score	HD
Noise	0.713	23.855
Logits	0.830	5.424

(b)Dice loss

Dice Loss	Dice Score	HD
N	0.812	5.463
Y	0.830	5.424

(c)#steps (training)

Steps	Dice Score	HD
1000	0.821	5.223
5	0.830	5.424

6   Discussion

In this work, we developed a novel denoising diffusion probabilistic model for 3D image multiclass segmentation. By recycling the model's predictions at previous time steps to replace ground truth during training, the method aligns diffusion training and segmentation evaluation, resulting in significant performance improvements compared to existing diffusion methods. Other techniques mitigating training and test inconsistency further improved the diffusion model's performance . However, the diffusion model did not outperform the non-diffusion-based segmentation models, which have long been well-established. We believe it is important to report this lack of superior performance in 3D medical image segmentation, especially when experiments are limited to the same compute budget. Future work could consider other diffusion models such as discrete diffusion and more memory-efficient implementation to enable more complex architectures. Although the presented experimental results primarily demonstrated methodological development, the fact that these were obtained on two large clinical data sets represents a promising step towards real-world applications. Localising multiple anatomical structures in prostate MR images is key to MR-targeted biopsy, radiotherapy and tissue-preserving focal treatment for patients with urological diseases, while abdominal organ outlines can be directly used in planning gastroenterological procedures and hepatic surgery.

Acknowledgement. This work was supported by the EPSRC grant (EP/T029404/1), the Wellcome/EPSRC Centre for Interventional and Surgical Sciences (203145Z/16/Z), the International Alliance for Cancer Early Detection, an alliance between Cancer Research UK (C28070/A30912, C73666/A31378), Canary Center at Stanford University, the University of Cambridge, OHSU Knight Cancer Institute, University College London and the University of Manchester, and Cloud TPUs from Google's TPU Research Cloud (TRC).

References

1. Ali, H., Murad, S., Shah, Z.: Spot the fake lungs: generating synthetic medical images using neural diffusion models. In: Longo, L., O'Reilly, R. (eds.) Artificial Intelligence and Cognitive Science. Communications in Computer and Information Science, vol. 1662. pp. 32–39. Springer, Cham (2023). https://doi.org/10.1007/978-3-031-26438-2_3

2. Amit, T., Nachmani, E., Shaharbany, T., Wolf, L.: SegDiff: image segmentation with diffusion probabilistic models. arXiv preprint: arXiv:2112.00390 (2021)

3. Baranchuk, D., Rubachev, I., Voynov, A., Khrulkov, V., Babenko, A.: Label-efficient semantic segmentation with diffusion models. arXiv preprint: arXiv:2112.03126 (2021)

4. Chen, T., Li, L., Saxena, S., Hinton, G., Fleet, D.J.: A generalist framework for panoptic segmentation of images and videos. arXiv preprint: arXiv:2210.06366 (2022)

5. Dhariwal, P., Nichol, A.: Diffusion models beat GANs on image synthesis. In: Advances in Neural Information Processing Systems, vol. 34, pp. 8780–8794 (2021)

6. Fu, Y., et al.: More unlabelled data or label more data? A study on semi-supervised laparoscopic image segmentation. arXiv preprint: arXiv:1908.08035 (2019)

7. Ho, J., Jain, A., Abbeel, P.: Denoising diffusion probabilistic models. In: Advances in Neural Information Processing Systems, vol. 33, pp. 6840–6851 (2020)

8. Ho, J., Salimans, T.: Classifier-free diffusion guidance. arXiv preprint: arXiv:2207.12598 (2022)

9. Ji, Y., et al.: AMOS: a large-scale abdominal multi-organ benchmark for versatile medical image segmentation. arXiv preprint: arXiv:2206.08023 (2022)

10. Khader, F., et al.: Medical diffusion-denoising diffusion probabilistic models for 3D medical image generation. arXiv preprint: arXiv:2211.03364 (2022)

11. Kim, B., Ye, J.C.: Diffusion deformable model for 4D temporal medical image generation. In: Wang, L., Dou, Q., Fletcher, P.T., Speidel, S., Li, S. (eds.) Medical Image Computing and Computer Assisted Intervention – MICCAI 2022. Lecture Notes in Computer Science, vol. 13431, pp. 539–548. Springer, Cham (2022). https://doi.org/10.1007/978-3-031-16431-6_51

12. Kingma, D., Salimans, T., Poole, B., Ho, J.: Variational diffusion models. In: Advances in Neural Information Processing Systems, vol. 34, pp. 21696–21707 (2021)

13. Kolbeinsson, B., Mikolajczyk, K.: Multi-class segmentation from aerial views using recursive noise diffusion. arXiv preprint: arXiv:2212.00787 (2022)
14. Lai, Z., et al.: Denoising diffusion semantic segmentation with mask prior modeling. arXiv preprint: arXiv:2306.01721 (2023)
15. Li, X., et al.: BrainGNN: interpretable brain graph neural network for fMRI analysis. Med. Image Anal. **74**, 102233 (2021)
16. Li, Y., et al.: Prototypical few-shot segmentation for cross-institution male pelvic structures with spatial registration. arXiv preprint: arXiv:2209.05160 (2022)
17. Nichol, A.Q., Dhariwal, P.: Improved denoising diffusion probabilistic models. In: International Conference on Machine Learning, pp. 8162–8171. PMLR (2021)
18. Pinaya, W.H., et al.: Brain imaging generation with latent diffusion models. arXiv preprint: arXiv:2209.07162 (2022)
19. Radford, A., et al.: Learning transferable visual models from natural language supervision. In: International Conference on Machine Learning, pp. 8748–8763. PMLR (2021)
20. Rombach, R., Blattmann, A., Lorenz, D., Esser, P., Ommer, B.: High-resolution image synthesis with latent diffusion models. In: Proceedings of the IEEE/CVF Conference on Computer Vision and Pattern Recognition, pp. 10684–10695 (2022)
21. Ronneberger, O., Fischer, P., Brox, T.: U-net: convolutional networks for biomedical image segmentation. In: Navab, N., Hornegger, J., Wells, W., Frangi, A. (eds.) Medical Image Computing and Computer-Assisted Intervention - MICCAI 2015. Lecture Notes in Computer Science(), vol. 9351, pp. 234–241. Springer, Cham (2015). https://doi.org/10.1007/978-3-319-24574-4_28
22. Saeed, S.U., et al.: Bi-parametric prostate MR image synthesis using pathology and sequence-conditioned stable diffusion. arXiv preprint: arXiv:2303.02094 (2023)
23. Sohl-Dickstein, J., Weiss, E., Maheswaranathan, N., Ganguli, S.: Deep unsupervised learning using nonequilibrium thermodynamics. In: International Conference on Machine Learning, pp. 2256–2265. PMLR (2015)
24. Song, Y., Dhariwal, P., Chen, M., Sutskever, I.: Consistency models. arXiv preprint: arXiv:2303.01469 (2023)
25. Strudel, R., Garcia, R., Laptev, I., Schmid, C.: Segmenter: transformer for semantic segmentation. In: Proceedings of the IEEE/CVF International Conference on Computer Vision, pp. 7262–7272 (2021)
26. Wolleb, J., Sandkühler, R., Bieder, F., Valmaggia, P., Cattin, P.C.: Diffusion models for implicit image segmentation ensembles. In: International Conference on Medical Imaging with Deep Learning, pp. 1336–1348. PMLR (2022)
27. Wu, J., Fang, H., Zhang, Y., Yang, Y., Xu, Y.: MedSegDiff: medical image segmentation with diffusion probabilistic model. arXiv preprint: arXiv:2211.00611 (2022)
28. Young, S.I., Dalca, A.V., Ferrante, E., Golland, P., Fischl, B., Iglesias, J.E.: Sud: supervision by denoising for medical image segmentation. arXiv preprint: arXiv:2202.02952 (2022)

Applications

Diffusion-Based Data Augmentation for Skin Disease Classification: Impact Across Original Medical Datasets to Fully Synthetic Images

Mohamed Akrout[1]([✉]), Bálint Gyepesi[1], Péter Holló[2], Adrienn Poór[2],
Blága Kincső[2], Stephen Solis[1], Katrina Cirone[1], Jeremy Kawahara[1],
Dekker Slade[1], Latif Abid[1], Máté Kovács[1], and István Fazekas[1]

[1] AIP Labs, Budapest, Hungary
mohamed@aip.ai
[2] Faculty of Medicine, Department of Dermatology, Venereology and
Dermatooncology, Semmelweis University, Budapest, Hungary

Abstract. Despite continued advancement in recent years, deep neural networks still rely on large amounts of training data to avoid overfitting. However, labeled training data for real-world applications such as healthcare is limited and difficult to access given longstanding privacy, and strict data sharing policies. By manipulating image datasets in the pixel or feature space, existing data augmentation techniques represent one of the effective ways to improve the quantity and diversity of training data. Here, we look to advance augmentation techniques by building upon the emerging success of text-to-image diffusion probabilistic models in augmenting the training samples of our macroscopic skin disease dataset. We do so by enabling fine-grained control of the image generation process via input text prompts. We demonstrate that this generative data augmentation approach successfully maintains a similar classification accuracy of the visual classifier even when trained on a fully synthetic skin disease dataset. Similar to recent applications of generative models, our study suggests that diffusion models are indeed effective in generating high-quality skin images that do not sacrifice the classifier performance, and can improve the augmentation of training datasets after curation.

Keywords: Data augmentation · Skin condition classification · AI for dermatology · Diffusion models · Synthetic medical datasets

1 Introduction

The last months have witnessed the emergence of diffusion probabilistic models (DPM) [10] as a powerful generator of high-fidelity synthetic datasets, leading

---

M. Akrout and B. Gyepesi—Equal contribution.

© The Author(s), under exclusive license to Springer Nature Switzerland AG 2024
A. Mukhopadhyay et al. (Eds.): DGM4MICCAI 2023, LNCS 14533, pp. 99–109, 2024.
https://doi.org/10.1007/978-3-031-53767-7_10

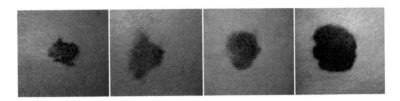

Fig. 1. Synthetic melanoma images generated by the stable diffusion model after fine-tuning it with melanoma images using the input text prompt "melanoma".

to record-breaking performances in various applications such as image synthesis [21], natural language processing [4], and computational chemistry [3], to name a few. When compared to other types of generative models, such as generative adversarial networks (GANs) and variational autoencoders, DPMs are easier to train and offer state-of-the-art image generation quality [7]. Given that synthetic images play a crucial role in privacy-preserving generation and small dataset augmentation, DPMs attracted significant attention in the medical imaging field. Table 1 provides an overview of the prior studies of DPMs, including their medical applications and dataset domains. At first glance, the reader can identify that the study in [23] is the closest one to this work where synthetic images were generated from seed images in the Fitzpatrick 17k dataset using the OpenAI's DALL·E 2 model [19].

Table 1. Summary of existing applications of diffusion models in medical imaging.

Medical applications	Dataset domain	Papers
Image generation	lungs X-Ray, CT, MRI	[2,5,16,17]
Image segmentation	MRI, CT, ultrasound	[9,13,30]
Image inpainting	MRI	[22]
Image denoising	MRI, CT, retinal OCT	[6,11,32]
Lesion detection	MRI	[24,29,31]
Image translation	MRI, CT	[13,15]
Seed-image based augmentation	Dermatology	[23]
Skin disease classification using large synthetic datasets	**Dermatology**	**This work**

Inspired by the recent early success of DPMs, we propose to use diffusion models for image augmentation as part of supervised machine learning pipelines. More specifically, we study how diffusion models can i) increase the classification metrics for skin diseases, and ii) augment skin condition datasets by effectively manipulating the generated images' features conditioned on the input text prompts. This paper makes the following contributions:

- We study the potential of DPMs for skin disease classifications by fine-tuning them on six different disease conditions: basal cell carcinoma, melanoma, actinic keratosis, atypical melanocytic nevus, lentigo, seborrheic keratosis. We do so by learning the embeddings of each disease using text inversion.

- We demonstrate that the classification accuracies of skin disease classifiers trained on generated synthetic images is similar to training on real images, where the performance is maintained when using half the number of real images, and only slightly deteriorates when using a fully synthetic dataset. This result suggests that the recent success of generative models can help minimize the barriers of sharing labeled medical datasets, with minimal performance deterioration.
- We illustrate how DPMs are powerful tools to add visual aspects of skin images guided by domain experts in complementing training datasets.

2   Diffusion-Based Data Augmentation

In this section, we begin by describing the methods used for training the embeddings of the aforementioned six skin diseases on our macroscopic skin images. Then, we present the datasets associated with the two DPM training scenarios: a hybrid dataset compromising 50% synthetic and 50% real images, and a 100% fully synthetic dataset generated by the trained embeddings.

2.1   Stable Diffusion

The stable diffusion model proposed in [21] is not a monolithic model, but rather a pipeline of three components, as depicted in Fig. 2:

1) *Text encoding*, based on the CLIP model [18], which transforms each token of the input text prompt into an embedding vector.
2) *Latent space U-Net generator*, which takes all the token embeddings and a random noise array (a.k.a., latent array) and sequentially generates multiple arrays that better resemble the input text and the visual images on which the U-Net has been trained.
3) *Image decoder*, based on a variational autoencoder (VAE) to transform the obtained latent array into the pixel space.

In this pipeline, the embedding vectors of the text encoding control both the generation of the U-Net latent space representations as well as the VAE decoding.

2.2   Training Dataset for Synthetic Image Generation

The limited number of available labeled images is one of the leading limitations faced by medical classification applications. Our internal macroscopic image dataset consists of thousands of skin condition images curated and classified by dermatologists to cover more than 700 different diseases. Here, we choose six widely spread classes across three distinct categories:

- *Malignant classes*: basal cell carcinoma and melanoma;
- *Pre-malignant classes*: actinic keratosis and atypical melanocytic nevus;
- *Benign classes*: lentigo and seborrheic keratosis.

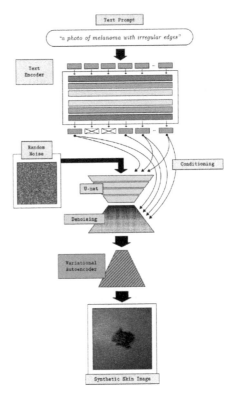

Fig. 2. The diffusion model pipeline for synthetic skin image generation.

Table 2 provides an overview of the number of images used for each disease in training the text embedding with the stable diffusion model.

In order to train the text embeddings associated to each skin disease, we use the stable diffusion architecture [20] based on latent diffusion models [21]. Using a model of the latter pretrained on multiple LAION datasets [1], we fine-tune each embedding on our real-world image skin condition dataset for two million steps using the default hyperparameters proposed in [25]. We use PyTorch for both training and inference. Each embedding is trained on three NVIDIA GeForce RTX 3090 GPUs.

Table 2. The number of real training images for the considered skin diseases.

Category	Skin disease	Data source
Benign	Seborrheic keratosis	2134
	Lentigo	680
Pre-malignant	Actinic keratosis	3298
	Atypical melanocytic nevus	623
Malignant	Basal cell carcinoma	7081
	Melanoma	3381

2.3   Curation of Generated Images

While most of the generated skin disease images are of high quality, it is not unusual to obtain generated images of medium or low quality. To isolate high-quality images from lower qualities, Fig. 3 depicts the full pipeline for augmenting our skin disease dataset composed of the following four steps:

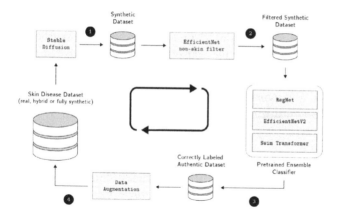

Fig. 3. Summary of the four steps of the generation pipeline for skin disease data augmentation.

1) *Synthetic data generation*: Using the stable diffusion model described in Sect. 2.1, we generate 30.000 images for each one of the considered six skin diseases to get a synthetic dataset.
2) *Non-skin image filtering*: We run the obtained synthetic dataset in 1) through a pretrained binary EfficientNet classifier [26] to filter out any non-skin images. The binary classifier has been trained on the skin images of the macroscopic dataset presented in Table 2 and non-skin images from ImageNet. The accepted images as skin images by the binary classifier represent more than 99% of the generated images and constitute the filtered synthetic dataset.

3) *Skin disease image filtering*: We use the filtered synthetic dataset to predict the skin disease label using a pretrained ensemble model composed of two CNN models (EfficientNetV2 [27], RegNet [8]) and a visual transformer (Swin-Transformer [14]). This ensemble model has been pretrained on the macroscopic dataset presented in Table 2.

4) *Data augmentation*: We use the correctly labeled images by the pretrained ensemble classifier as the data source for augmenting our initial dataset.

3   Experiments and Results

3.1   Dataset Scenarios for Synthetic Image Generation

Based on the filtered images whose labels were correctly predicted by the pretrained ensemble classifier, we build a fully synthetic dataset consisting of 500 images per skin disease. For the real images, we randomly sample 500 images per class from our macroscopic skin image dataset. To examine the impact of the synthetic dataset on classification metrics, we consider the following datasets:

- a *small real dataset* (real-small) containing 250 real images only,
- a *real dataset* containing 500 real images only,
- a *hybrid dataset* consisting of 250 real images and 250 synthetic images,
- a *synthetic dataset* containing 500 synthetic images only.

Note that the four datasets are balanced across skin diseases with varying proportions of real and synthetic images. This allows us to assess the efficiency of substituting real data with synthetic ones.

3.2   Medical Synthetic Data Samples Using Text Prompt Inputs

Here, we demonstrate the quality of the synthetic skin disease images stemming from the generation pipeline in Fig. 3 by providing four synthetic images for each disease. Similar to the synthetic melanoma images in Fig. 1, we present synthetic images of seborrheic keratosis, lentigo, atypical melanocytic nevus, basal cell carcinoma and actinic keratosis in Figs. 4, 5, 6, and 7, respectively.

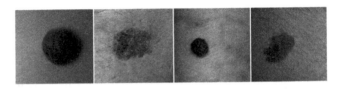

Fig. 4. Synthetic seborrheic keratosis images generated by the stable diffusion model after fine-tuning it with seborrheic keratosis images using the input text prompt "seborrheic keratosis".

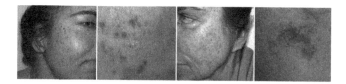

Fig. 5. Synthetic lentigo images generated by the stable diffusion model after fine-tuning it with lentigo images using the input text prompt "lentigo".

While the impressive generative capabilities of AI models have already been established for normal and glaucomatous eyes in [12], our generated macroscopic images for different skin diseases similarly establishes the effectiveness for dermatology using larger synthetic datasets. This is to be opposed to seed-image based augmentation in [23] where synthetic datasets where not used to fine-tune the generative model.

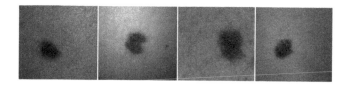

Fig. 6. Synthetic synthetic atypical melanocytic nevus images generated by the stable diffusion model after fine-tuning it with atypical melanocytic nevus images using the input text prompt "atypical melanocytic nevus".

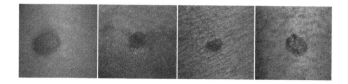

Fig. 7. Synthetic basal cell carcinoma images generated by the stable diffusion model after fine-tuning it with basal cell carcinoma images using the input text prompt "basal cell carcinoma".

3.3   Classification of Skin Conditions

In this section, we first describe the training and inference procedures of the skin disease ensemble classifier on the four datasets described in Sect. 3.1.

The Training Step. We start by training three networks of the ensemble classifier (i.e., Swin-Transformer [14], EfficientNetV2 [27], and RegNetZ [8]) on each one of the datasets (i.e., real, hybrid, and synthetic). We do so using the PyTorch Image Models library [28]. We make use of the default training hyperparameters and set the number of training epochs and batch size to 100 and 8, respectively. We also use early stopping[1] by monitoring the validation loss, and opt for the stochastic gradient descent (SGD) optimizer. We also use a data split of 80% and 20% for training and validation dataset sizes, respectively.

For every dataset, we calculate the mean and standard deviation for each one of the RBG image channels. They are accustomed to preprocessing the input images to normalize the images fed to all the networks. It is worth noting that the early stopping criterion occurs when we train the models on the fully synthetic dataset only. This is as opposed to training on real or hybrid datasets, where early stopping does not occur because the validation accuracy stagnates with very little increase, and peaks at 89% only. This observation suggests that the fully synthetic dataset generated with stable diffusion exhibits non-perceptible differentiating features that is allowing for faster training and convergence.

The Inference Step. We evaluate the trained ensemble model by running inference on our test dataset consisting of 3582 real images. Table 3 shows their distribution across the skin disease categories and classes.

Table 3. The number of test images for the six considered skin diseases

Category	Skin disease	Number of images
Benign	Seborrheic keratosis	1597
	Lentigo	293
Pre-malignant	Actinic keratosis	282
	Atypical melanocytic nevus	885
Malignant	Basal cell carcinoma	345
	Melanoma	180

We do not carry out any preprocessing to the test images other than the same normalization applied to the training images.

3.4   Classification Results

We now evaluate three ensemble classifiers where each classifier is separately trained on one of the real-small, real, hybrid and synthetic datasets, as described in Sect. 3.1. We run inference on our test dataset and report in Table 4 the

---

[1] Here, early stopping occurs as soon as the validation accuracy does not improve over 10 consecutive epochs.

associated top-k classification accuracy. The latter computes the number of times where the correct skin disease is among the top-k predicted diseases (ranked from highest to lowest predicted scores).

Table 4. Top-1 to top-5 skin disease classification accuracy on real-small, real, hybrid and fully synthetic datasets.

Dataset	# of images		Accuracy				
	Real	Synthetic	Top-1	Top-2	Top-3	Top-4	Top-5
Real-small	250	0	53.41%	73.51%	83.22%	89.75 %	95.45%
Real	500	0	54.05%	**73.95%**	84.84%	91.49 %	**96.96%**
Hybrid	250	250	**54.13%**	73.23%	**85.01%**	**92.16%**	96.65%
Synthetic	0	500	47.29%	70.71%	84.09%	**92.16%**	96.85%

From Table 4, it can be seen that the top-k accuracies of the four classifiers are very comparable. More importantly, we observe how the use of synthetic images improves the overall accuracy of skin classifiers. Indeed, their performances on the real and hybrid datasets have been improved. As ascertained by our clinical partners at Semmelweis University, this result confirms that beyond their impressive visual quality across thousands of images, diffusion models also provide significant benefit as synthetic images for real-world medical applications.

4 Conclusion

In this paper, we demonstrate the impressive generative capabilities of probabilistic diffusion models in generating macroscopic skin disease images. We show how it is possible to condition the probabilistic diffusion-based generation on text prompt inputs in obtaining fine-grained synthetic images. Furthermore, we propose a closed loop data augmentation pipeline to automatically curate the generated images while complementing real-world skin disease datasets. Finally, our classification task of six skin diseases highlights how synthetic images are reliable data sources given that they have been demonstrated beneficial for skin disease classification. This result underlines the importance of the recent generative modelling success for medical applications as an effective means of data sharing without infringing confidentiality issues. Several exciting avenues for further investigation remain open such as conditioning the image generation in relation to skin tone, with skin tone diversification in datasets being another leading limitation, or the use of input images in addition to the text prompt.

References

1. Large-scale Artificial Intelligence Open Network. https://laion.ai. Accessed 11 Jan 2023
2. Ali, H., Murad, S., Shah, Z.: Spot the fake lungs: generating synthetic medical images using neural diffusion models. arXiv preprint: arXiv:2211.00902 (2022)
3. Anand, N., Achim, T.: Protein structure and sequence generation with equivariant denoising diffusion probabilistic models. arXiv preprint: arXiv:2205.15019 (2022)
4. Austin, J., Johnson, D.D., Ho, J., Tarlow, D., van den Berg, R.: Structured denoising diffusion models in discrete state-spaces. In: Advances in Neural Information Processing Systems, vol. 34, pp. 17981–17993 (2021)
5. Chambon, P., Bluethgen, C., Langlotz, C.P., Chaudhari, A.: Adapting pretrained vision-language foundational models to medical imaging domains. arXiv preprint: arXiv:2210.04133 (2022)
6. Chung, H., Lee, E.S., Ye, J.C.: MR image denoising and super-resolution using regularized reverse diffusion. arXiv preprint: arXiv:2203.12621 (2022)
7. Croitoru, F.A., Hondru, V., Ionescu, R.T., Shah, M.: Diffusion models in vision: a survey. arXiv preprint: arXiv:2209.04747 (2022)
8. Dollár, P., Singh, M., Girshick, R.: Fast and accurate model scaling. FAIR (2021)
9. Guo, X., Yang, Y., Ye, C., Lu, S., Xiang, Y., Ma, T.: Accelerating diffusion models via pre-segmentation diffusion sampling for medical image segmentation. arXiv preprint: arXiv:2210.17408 (2022)
10. Ho, J., Jain, A., Abbeel, P.: Denoising diffusion probabilistic models. In: Advances in Neural Information Processing Systems, vol. 33, pp. 6840–6851 (2020)
11. Hu, D., Tao, Y.K., Oguz, I.: Unsupervised denoising of retinal OCT with diffusion probabilistic model. In: Medical Imaging 2022: Image Processing, vol. 12032, pp. 25–34. SPIE (2022)
12. Kumar, A.J.S., et al.: Evaluation of generative adversarial networks for high-resolution synthetic image generation of circumpapillary optical coherence tomography images for glaucoma. JAMA Ophthalmol. **140**(10), 974–981 (2022)
13. La Barbera, G., et al.: Anatomically constrained CT image translation for heterogeneous blood vessel segmentation. arXiv preprint: arXiv:2210.01713 (2022)
14. Liu, Z., et al.: Swin transformer: hierarchical vision transformer using shifted windows. Microsoft Research Asia (2021)
15. Özbey, M., et al.: Unsupervised medical image translation with adversarial diffusion models. arXiv preprint: arXiv:2207.08208 (2022)
16. Packhäuser, K., Folle, L., Thamm, F., Maier, A.: Generation of anonymous chest radiographs using latent diffusion models for training thoracic abnormality classification systems. arXiv preprint: arXiv:2211.01323 (2022)
17. Pinaya, W.H., et al.: Brain imaging generation with latent diffusion models. In: Mukhopadhyay, A., Oksuz, I., Engelhardt, S., Zhu, D., Yuan, Y. (eds.) Deep Generative Models. Lecture Notes in Computer Science, vol. 13609, pp. 117–126. Springer, Cham (2022). https://doi.org/10.1007/978-3-031-18576-2_12
18. Radford, A., et al.: Learning transferable visual models from natural language supervision. In: International Conference on Machine Learning, pp. 8748–8763. PMLR (2021)
19. Ramesh, A., Dhariwal, P., Nichol, A., Chu, C., Chen, M.: Hierarchical text-conditional image generation with clip Latents. arXiv preprint: arXiv:2204.06125 (2022)

20. Rombach, R., Blattman, A., Lorenz, D., Esser, P., Ommer, B.: Stable diffusion (2022). https://github.com/CompVis/stable-diffusion
21. Rombach, R., Blattmann, A., Lorenz, D., Esser, P., Ommer, B.: High-resolution image synthesis with latent diffusion models. In: Proceedings of the IEEE/CVF Conference on Computer Vision and Pattern Recognition, pp. 10684–10695 (2022)
22. Rouzrokh, P., Khosravi, B., Faghani, S., Moassefi, M., Vahdati, S., Erickson, B.J.: Multitask brain tumor inpainting with diffusion models: a methodological report. arXiv preprint: arXiv:2210.12113 (2022)
23. Sagers, L.W., Diao, J.A., Groh, M., Rajpurkar, P., Adamson, A.S., Manrai, A.K.: Improving dermatology classifiers across populations using images generated by large diffusion models. arXiv preprint: arXiv:2211.13352 (2022)
24. Sanchez, P., Kascenas, A., Liu, X., O'Neil, A.Q., Tsaftaris, S.A.: What is healthy? Generative counterfactual diffusion for lesion localization. In: Mukhopadhyay, A., Oksuz, I., Engelhardt, S., Zhu, D., Yuan, Y. (eds.) Deep Generative Models. Lecture Notes in Computer Science, vol. 13609, pp. 34–44. Springer, Cham (2022). https://doi.org/10.1007/978-3-031-18576-2_4
25. Stein, L.: Invoke AI (2022). https://github.com/invoke-ai/InvokeAI
26. Tan, M., V.Le, Q.: EfficientNet: rethinking model scaling for convolutional neural networks (2020)
27. Tan, M., Le, Q.: EfficientNetV2: smaller models and faster training (2021)
28. Wightmann, R.: PyTorch image models (2022). https://github.com/rwightman/pytorch-image-models
29. Wolleb, J., Bieder, F., Sandkühler, R., Cattin, P.C.: Diffusion models for medical anomaly detection. arXiv preprint: arXiv:2203.04306 (2022)
30. Wu, J., Fang, H., Zhang, Y., Yang, Y., Xu, Y.: MedSegDiff: medical image segmentation with diffusion probabilistic model. arXiv preprint: arXiv:2211.00611 (2022)
31. Wyatt, J., Leach, A., Schmon, S.M., Willcocks, C.G.: AnoDDPM: anomaly detection with denoising diffusion probabilistic models using simplex noise. In: Proceedings of the IEEE/CVF Conference on Computer Vision and Pattern Recognition, pp. 650–656 (2022)
32. Xia, W., Lyu, Q., Wang, G.: Low-dose CT using denoising diffusion probabilistic model for 20x speedup. arXiv preprint: arXiv:2209.15136 (2022)

Unsupervised Anomaly Detection in 3D Brain FDG PET: A Benchmark of 17 VAE-Based Approaches

Ravi Hassanaly[✉], Camille Brianceau, Olivier Colliot, and Ninon Burgos

Sorbonne Université, Institut du Cerveau - Paris Brain Institute - ICM, CNRS, Inria, Inserm, AP-HP, Hôpital de la Pitié Salpêtrière, 75013 Paris, France
`ravi.hassanaly@icm-institute.org`

Abstract. The use of deep generative models for unsupervised anomaly detection is an area of research that has gained interest in recent years in the field of medical imaging. Among all the existing models, the variational autoencoder (VAE) has proven to be efficient while remaining simple to use. Much research to improve the original method has been achieved in the computer vision literature, but rarely translated to medical imaging applications. To fill this gap, we propose a benchmark of fifteen variants of VAE that we compare with a vanilla autoencoder and VAE for a neuroimaging use case relying on a simulation-based evaluation framework. The use case is the detection of anomalies related to Alzheimer's disease and other dementias in 3D FDG PET.

We show that among the fifteen VAE variants tested, nine lead to a good reconstruction accuracy and are able to generate healthy-looking images. This indicates that many approaches developed for computer vision applications can generalize to the unsupervised detection of anomalies of various shapes, intensities and locations in 3D FDG PET. However, these models do not outperform the vanilla autoencoder and VAE.

Keywords: Variational autoencoder · Deep generative models · Unsupervised anomaly detection · PET · Alzheimer's disease

1 Introduction

Recent advances in medical image analysis have allowed the emergence of algorithms that can perform complex tasks such as computer-aided diagnosis [7,10] with pseudo-healthy reconstruction for unsupervised anomaly detection (UAD). Contrary to supervised approaches, UAD does not require human annotations that are costly and time consuming, and enables the detection of any type of anomalies, without having seen them before. Most approaches rely on generative models to reconstruct healthy looking images, also called pseudo-healthy images [1,7,10]. The assumption is that if a model is trained with images from subjects diagnosed as healthy, the reconstruction of images with a pathology should not contain

Supplementary Information The online version contains supplementary material available at https://doi.org/10.1007/978-3-031-53767-7_11.

pathology-specific features and look like a healthy image. Comparing the pseudo-healthy reconstruction with the real image then allows the detection of anomalies.

The application context of our work is the detection of metabolic changes visible in brain ^{18}F-fluorodeoxyglucose (FDG) positron emission tomography (PET) caused by Alzheimer's disease and other dementias [8]. These subtle changes appear several years before the first symptoms and can be used for early diagnosis [16]. In neuroimaging, deep learning methods for UAD have not been much applied for the diagnosis of dementia [9]. It is a challenging task because the metabolic abnormalities are diffuse and little intense, which makes them difficult to detect [3].

The different pseudo-healthy reconstruction approaches that have been developed for medical imaging rely on variational autoencoders (VAEs) [19], generative adversarial networks (GANs) [12] and more recently diffusion models [15]. We aim to compare VAE-based models as they have shown their efficacy for UAD in medical imaging [1,7], are easy to train, easily scalable, with good interpretation capacity thanks to their regularized latent space, and are able to handle small datasets. Much research to improve the original VAE has been achieved in the computer vision literature [2,5,11,14,18,21–23,25,27,29,30,32,36], but only a few have been translated to medical imaging applications [1,6,9,24,31].

We propose a benchmark of seventeen VAE-based models and show results in the context of pseudo-healthy reconstruction for dementia from 3D FDG PET. As far as we know, the only study that has compared VAEs for neuroimaging data is that of Baur et al. [1]. However, it was restricted to models that had already been used for medical imaging applications. Many other VAE extensions have thus not been assessed. Also, it was dedicated to the detection of very sharp and intense anomalies, such as brain tumors or multiple sclerosis lesions, which is very different from the identification of subtle anomalies found in PET images of patients with cognitive disorders. Finally, it was performed in 2D. Our work aims to contribute to this effort by evaluating a much wider set of approaches, including many that were never used in medical imaging, relying on the work of Chadebec et al. [4]. This will provide an insight into the performance that such models can achieve in detecting anomalies in 3D data when trained with a relatively small dataset (few hundreds of images) compared to most datasets used in the computer vision literature (several tens of thousands images). The models will be evaluated and compared based on reconstruction quality and on their ability to generate healthy looking images using a previously proposed simulation framework [13].

2   Methods

2.1   Variational Autoencoder Framework for Pseudo-healthy Image Reconstruction

Let D be a set of medical images of the same modality acquired following a similar protocol. D can contain healthy and pathological images and can be divided in respectively two complementary subsets D_h and D_p. Let's take as an example a set of FDG PET images $\mathbf{x} \in D_h$ whose distribution is $p(\mathbf{x})$.

The goal of pseudo-healthy image reconstruction is to generate an FDG PET image of healthy appearance. The idea is to approximate the healthy image true distribution $p(\mathbf{x})$ with a chosen model $p_\theta(\mathbf{x})$ such that $p_\theta(\mathbf{x}) \approx p(\mathbf{x})$. Then, during reconstruction, the images (of healthy subjects or patients) are projected into that "healthy images" learned subspace by the generative model.

This can be modeled using the VAE framework [19] by assuming that a latent variable $\mathbf{z}$ is involved in the generation process of $\mathbf{x}$: $p_\theta(\mathbf{x}) = \int_z p(\mathbf{z})p_\theta(\mathbf{x} \mid \mathbf{z})\mathrm{d}\mathbf{z}$ where $\mathbf{z} \sim p_\theta(\mathbf{z})$ is the prior distribution on the latent space and $p_\theta(\mathbf{x} \mid \mathbf{z})$ is the generative model (or the decoder) that learns to generate healthy images from $\mathbf{z}$. To compute the appropriate $\mathbf{z}$ for each data input $\mathbf{x}$ of our dataset, we need the posterior distribution $p_\theta(\mathbf{z} \mid \mathbf{x})$. Since it is untractable, we approximate it using variational inference by introducing another model $q_\phi(\mathbf{z} \mid \mathbf{x})$ such that $q_\phi(\mathbf{z} \mid \mathbf{x}) \approx p_\theta(\mathbf{z} \mid \mathbf{x})$. $q_\phi(\mathbf{z} \mid \mathbf{x})$ is the inference model (or encoder). Both the decoder and encoder are parametric models whose parameters are given by a neural network.

The objective is to maximize the likelihood of $p_\theta(\mathbf{x})$, which is equivalent to maximizing the evidence lower bound, which defines our loss function $\mathcal{L}_{\theta,\Phi}$ [20]

$$\log\left(p_\theta(\mathbf{x})\right) \geq \mathcal{L}_{\theta,\Phi}(x) = \mathbb{E}_{q_\Phi(\mathbf{z}\mid\mathbf{x})}\left[\log\left(p_\theta(\mathbf{x} \mid \mathbf{z})\right)\right] - D_{\mathrm{KL}}\left(q_\Phi(\mathbf{z} \mid \mathbf{x})\|p_\theta(\mathbf{z})\right) \quad (1)$$

with D_{KL} the Kullback-Leibler divergence.

During the training process, we learn an approximation of the posterior distribution $q_\phi(\mathbf{z} \mid \mathbf{x})$ for $x \in D_h$ as we train our model using only healthy subjects. When using the model for inference, we use this approximate posterior to estimate the latent variable $\mathbf{z}$ for $\mathbf{x} \in D$ (it can be from D_h or D_p).

2.2   Extensions to the Variational Autoencoder Framework

As explained in detail in [4], several contributions have been proposed to improve the VAE framework. They can be divided into four categories that correspond to different objectives:

- improve the prior distribution $p(\mathbf{z})$ by using a variational mixture of posteriors as prior (VAMP) [30], by learning the prior on a discrete latent space with vector quantized-VAE (VQVAE) [32], or by substituting the prior with a density estimation method using regularization with a gradient penalty (RAE-GP), or an ℓ^2 penalty on the decoder (RAE-ℓ^2) [11];
- better estimate the lower bound by using importance weighting (IWAE) [2], and using a linear normalizing flow (VAE-lin-NF) [25] or an inverse autoregressive flow (VAE-IAF) [21] to better estimate the posterior;
- encourage disentanglement of the features in the latent space by adding a weight to balance the terms of the loss in Eq. 1 (β -VAE) [14], decomposing the loss to show a total correlation term (β -TC VAE) [5], or by encouraging the distribution of the latent variable $q(\mathbf{z})$ to be factorial (FactorVAE) [18];

– and change the distance computed between the distributions by adding the mutual information between $\mathbf{x}$ and $\mathbf{z}$ as regularization (InfoVAE) [36], using another divergence term in the loss such as the maximum mean discrepancy in the Wasserstein autoencoder (WAE) [29] or a discriminator to differentiate a prior's sample from a posterior's sample in the adversarial autoencoder (AAE) [23], or by changing the reconstruction metric for another similarity metric such as the multi-scale structural similarity (MSSSIM-VAE) [27], or for the prediction of a discriminator on the output of the VAE (VAEGAN) [22].

In our benchmark, these models will be compared to the autoencoder (AE) and VAE [19], which makes a total of seventeen models. All of these methods have shown great results in other fields of computer vision, and, since VAE-based models can learn the data distribution on a small dataset, we keep the focus on them and aim to assess their performance in the context of medical imaging.

2.3   Evaluation of the Models

We can distinguish two main objectives when generating pseudo-healthy images: preserving the subject's identity in the reconstructed image and ensuring that the reconstruction appears healthy [35].

For the subject identity preservation, we evaluate the models on real images from healthy subjects only: the pseudo-healthy reconstruction of an image of a healthy subject should be identical to the input. This is assessed using three commonly used paired reconstruction metrics: the mean-squared error (MSE), the peak signal-to-noise ratio (PSNR) and the structural similarity (SSIM) [33].

To evaluate the capability of each model to reconstruct healthy looking images, since we do not have access to ground-truth lesions masks, we use the evaluation framework that has been introduced in [13]. It consist in simulating the effect of the disease by reducing the intensity of the PET uptake within regions associated with different dementias, thus mimicking regional hypometabolism [3]. After locally reducing the intensity of the image by a certain percentage, a Gaussian smoothing is applied to have a realistic result and diffuse anomalies. That way we can have pairs of diseased images with the original healthy scan that is used as ground-truth for the pseudo-healthy reconstruction as we do not have ground truths for images from real patients in our dataset. We simulate five different dementias on images of healthy subjects: Alzheimer's disease (AD), behavioral variant frontotemporal dementia (bvFTD), logopenic variant primary progressive aphasia (lvPPA), semantic variant PPA (svPPA) and posterior cortical atrophy (PCA). This allows us to evaluate the capability of the model to generalize to anomalies caused by different dementia subtypes. In addition, we simulate different degrees of AD severity by varying the reduction in intensity from five to seventy percents to study the sensitivity of the UAD approaches on subtle and severe anomalies. We compute the reconstruction error in the whole image, in the region associated with the simulated dementia and in the complementary of this region in the brain.

2.4   Materials

FDG PET scans used in this study were obtained from the publicly available ADNI database [17] (https://adni.loni.usc.edu). We selected FDG PET images co-registered, averaged and uniformized to a resolution of 8 mm FWHM to reduce the variability due to the use of different scanners. The images were then linearly registered to the standard MNI space, normalized in intensity using the average PET uptake in a region comprising cerebellum and pons, and cropped using the Clinica [26] `pet-linear` pipeline. We finally down-sampled the images to a voxel size of $80 \times 96 \times 80$ to reduce their dimension and the memory usage.

ADNI includes a total of 733 FDG PET scans of cognitively normal (CN) participants with a stable diagnosis over a three-year window (corresponding to 301 subjects). We discarded 144 images that were not correctly registered according to the quality check algorithms implemented in ClinicaDL [28].

2.5   Experimental Setting

We split our dataset of 247 remaining CN subjects at the subject's level to avoid data leakage [34]: 50 CN subjects (50 images) compose the test set, 19 subjects (19 images) belong to the validation set and 178 subjects (452 images) are used to train our models. The split is stratified by sex and age to reduce biases. The 50 images of the CN subjects from the test set are also used to simulate the hypometabolic images mimicking various dementias and AD severity degrees.

For the comparison to be as fair as possible, all the models share the same encoder and decoder architecture. The encoder is composed of three blocks that are the succession of a 3D convolutional layer and a batch normalization with a ReLU activation. Then the tensor is flatten and passes through a dense layer to output a one dimensional latent space. The decoder is almost symmetrical: it is composed of a dense layer followed by three blocks that are composed of a 3D deconvolutional layer and a batch normalization with a leaky ReLU activation. We tested several sizes of latent space (16, 64, 128 and 256), but as we observed similar performance, we report the results for a size of 128, consistent with the choice made in [1].

We also use the same training parameters and environment to train all the models. We trained each model on 300 epochs with a learning rate of 10^{-5} and a batch size of 24 on a HPC with Nvidia Tesla V100 GPUs that have 32GB of memory. We are aware that model performance can greatly vary depending on these parameters, but for fair comparison we decided to choose the best parameters on the VAE and use the same for all models. It takes on average between 1' and 1'30" to train one epoch with comparable performance for each model on our computer cluster, meaning around 7 h per model for 300 epochs.

VAE-based model implementation relies on Pythae [4] and neuroimage processing on ClinicaDL [28], two open source software tools. The code used for this study is available on GitHub and can be used to reproduce the experiments: https://github.com/ravih18/VAE-models-for-UAD.

3 Results

3.1 Pseudo-healthy Reconstruction from Images of Control Subjects

We first assessed whether the different models could preserve the subject's identity by computing the MSE, PSNR and SSIM between the input and reconstructed images of the CN subjects. Results are reported in Table 1. We observe that no model clearly outperforms the others. On the other hand, VAMP [30], VAE-lin-NF [25], MSSSIM-VAE [27] and VAEGAN [22] perform less well than the others (MSE > 0.05, PSNR < 20 dB, SSIM < 0.5). A possible explanation is that the dataset is too small for these models to learn the data distribution.

Table 1. Reconstruction metrics computed between the pseudo-healthy reconstructions obtained with the various models evaluated and the original healthy PET image of CN subjects from the test set. Light gray highlights the worst performing models.

Model	MSE ↓	PSNR (dB) ↑	SSIM ↑
AE	0.02694 ± 0.00603	25.78 ± 0.84	0.725 ± 0.033
VAE [19]	0.02471 ± 0.00517	26.15 ± 0.79	0.771 ± 0.027
VAMP [30]	1.09029 ± 0.10416	9.64 ± 0.41	0.057 ± 0.015
RAE-GP [11]	0.02363 ± 0.00480	26.34 ± 0.79	0.750 ± 0.030
RAE-ℓ^2 [11]	0.02385 ± 0.00532	26.31 ± 0.83	0.761 ± 0.029
VQVAE [32]	0.02645 ± 0.00608	25.87 ± 0.85	0.731 ± 0.032
IWAE [2]	0.03531 ± 0.00711	24.60 ± 0.80	0.692 ± 0.030
VAE-lin-NF [25]	0.12887 ± 0.02875	18.99 ± 0.89	0.483 ± 0.036
VAE-IAF [21]	0.02900 ± 0.00560	25.45 ± 0.77	0.706 ± 0.032
β -VAE [14]	0.03927 ± 0.00654	24.12 ± 0.71	0.708 ± 0.028
β -TC VAE [5]	0.02819 ± 0.00499	25.55 ± 0.67	0.729 ± 0.031
FactorVAE [18]	0.02869 ± 0.00550	25.49 ± 0.74	0.704 ± 0.032
InfoVAE [36]	0.03223 ± 0.00566	24.97 ± 0.69	0.706 ± 0.030
WAE [29]	0.02920 ± 0.00509	25.40 ± 0.66	0.690 ± 0.032
AAE [23]	0.02919 ± 0.00597	25.43 ± 0.81	0.709 ± 0.032
MSSSIM-VAE [27]	1.22541 ± 0.18918	9.17 ± 0.73	0.167 ± 0.027
VAEGAN [22]	0.86575 ± 0.03080	10.63 ± 0.15	0.073 ± 0.014

The other models obtain a similar performance with, on average, an MSE < 0.04, PSNR > 24 dB and SSIM comprised between 0.69 and 0.75. Not surprisingly, the AE leads to a good performance for this reconstruction task according to the MSE as it is the optimized metric. The vanilla VAE [19] seems to be one of the best models but does not stand out from the other models. It is probable that some models would benefit from hyper-parameter fine tuning to perform better, but it is interesting to see that optimal parameters obtained on classic computer vision datasets do generalize to this different application for many models.

3.2 Pseudo-healthy Reconstruction from Images Simulating Dementia

In the following, we discarded the four models that did not give acceptable reconstructions. We first report, for the five dementia subtypes considered simulated with a hypometabolism of 30%, the MSE and SSIM between the simulated image and their reconstructions within the binary mask where hypometabolism was applied (e.g. between X' and $\widehat{X'}$ within the binarized mask M in Fig. 1). All the models reach a very similar performance with an MSE on average across models of 0.0132 (min MSE of 0.0096 for the RAE GP [11] and max MSE of 0.0183 for the IWAE [2]) and an average SSIM of 0.710 (min SSIM of 0.684 for the IWAE [2] and max SSIM of 0.733 for the RAE-ℓ^2 [11]). This means that the VAE-based models can generalize to various kinds of anomalies located in different parts of the brain, and that none of the tested models can be selected based on this criteria. The average MSE over all the models and all the dementia subtypes (between X' and $\widehat{X'}$) is 0.0132 in the pathological masks M against 0.0072 outside the masks, which makes a 58.6% difference between both regions. The average SSIM is 0.710 inside masks M against 0.772 outside the masks for a 8.4% difference. This shows that the reconstruction error is much larger in regions that have been used for hypometabolism simulation, as expected. For comparison, the percentage difference is only 10.2% for the MSE and 0.2% for the SSIM when computed between the pseudo-healthy reconstruction $\widehat{X'}$ and the real pathology-free images X. This illustrates that the models are all capable of reconstructing the pathological regions as healthy.

We then report in Fig. 2 the MSE within the mask simulating AD when generating hypometabolism of various degrees (5% to 70%) for each model. It is interesting to observe that most of the models could be used to detect anomalies of higher intensity as they have an increasing difference in terms of MSE for hypometabolism of 20% and more. The same trend was observed with the SSIM. The RAE-ℓ^2 [11] does not scale as well as other models, probably because the

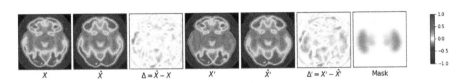

Fig. 1. Example of FDG PET image of a CN subject (X) with the corresponding pseudo-healthy reconstruction $(\widehat{X})$ and difference image (Δ), followed by an image simulating AD hypometabolism obtained from X (X') with the corresponding pseudo-healthy reconstruction $(\widehat{X'})$ and difference image (Δ'), and the mask used to generate X' (M). The pseudo-healthy reconstructions were obtained from the vanilla VAE model.

regularization is done on the decoder weights so nothing prevents the encoder from learning a posterior that is less general. We also notice that the IWAE [2] has a worse reconstruction on the pathological region compared to other models, and this becomes more pronounced when the severity of the disease is increased. However this does not mean that IWAE [2] better detects pathological areas since the reconstruction is poor in the whole image as well, meaning that IWAE [2] cannot perform well when the image is out of the training distribution. Surprisingly, the simple autoencoder gives similar results as other methods.

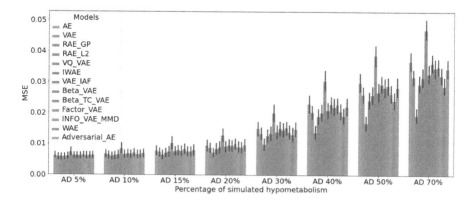

Fig. 2. Bar plot of the evolution of the MSE when computed within the mask characteristic of AD between the image simulated with different degrees of hypometabolism and its reconstruction. We observe that most models can scale to large anomalies.

4   Conclusion

The proposed benchmark aimed to introduce the use of recent VAE variants with medical imaging data of high dimension and compare their performance on the detection of dementia-related anomalies on 3D FDG PET brain images. We observed that most models have a comparable reconstruction ability when fed with images of healthy subjects and that their outputs correspond to healthy looking images when fed with images simulating anomalies. Exceptions are the VAEGAN [22], VAMP [30], VAE-lin-NF [25], MSSSIM-VAE [27], RAE-ℓ^2 [11] and IWAE [2]. Thanks to the evaluation framework that consists in simulating images with anomalies from pathology-free images, we showed that most models can generalize pseudo-healthy reconstruction to different dementias and different severity degrees. These results are interesting as it means that VAE-based models developed for natural images can generalize well to other tasks (here 3D brain imaging): they are easy to use and do not necessarily require a large training set, which might not be the case for other types of generative models. We also showed that in our scenario (small dataset of complex 3D images) the simplest models (vanilla AE and VAE) lead to results comparable to that of the more complex ones. Nevertheless, the results are for now limited to the detection of simulated anomalies. An evaluation on real images would be necessary to confirm these observations.

The proposed benchmark could be used in future work to assess whether the posterior learned by the different models is the same for images from healthy and diseased subjects using the simulation framework to compare the latent representation of both the original and simulated images, thus explaining the results of the models. It would also be interesting to compare some of the VAE-based models to GANs or diffusion models, and assess whether it would be possible to improve reconstruction quality while learning the distribution of healthy subject images.

Acknowledgment. The research leading to these results has received funding from the French government under management of Agence Nationale de la Recherche as part of the "Investissements d'avenir" program, reference ANR-19-P3IA-0001 (PRAIRIE 3IA Institute) and reference ANR-10-IAIHU-06 (Agence Nationale de la Recherche-10-IA Institut Hospitalo-Universitaire-6). This work was granted access to the HPC resources of IDRIS under the allocation AD011011648 made by GENCI.

Data used in preparation of this article were obtained from the Alzheimer's Disease Neuroimaging Initiative (ADNI) database (adni.loni.usc.edu).

References

1. Baur, C., Denner, S., Wiestler, B., Navab, N., Albarqouni, S.: Autoencoders for unsupervised anomaly segmentation in brain MR images: a comparative study. Media **69**, 101952 (2021)
2. Burda, Y., Grosse, R.B., Salakhutdinov, R.: Importance weighted autoencoders. In: ICLR (2016)
3. Burgos, N., et al.: Anomaly detection for the individual analysis of brain PET images. J. Med. Imag. **8**(2), 024003 (2021)
4. Chadebec, C., Vincent, L.J., Allassonniere, S.: Pythae: unifying generative autoencoders in Python - a benchmarking use case. In: Thirty-sixth Conference on NeurIPS Datasets and Benchmarks Track (2022)
5. Chen, R.T., Li, X., Grosse, R.B., Duvenaud, D.K.: Isolating sources of disentanglement in variational autoencoders. In: Advances in NeurIPS, vol. 31 (2018)
6. Chen, X., Konukoglu, E.: Unsupervised detection of lesions in brain MRI using constrained adversarial auto-encoders. In: MIDL (2018)
7. Chen, X., Konukoglu, E.: Unsupervised abnormality detection in medical images with deep generative methods, pp. 303–324. Elsevier (2022)
8. Chételat, G., et al.: Amyloid-PET and 18F-FDG-PET in the diagnostic investigation of Alzheimer's disease and other dementias. Lancet Neurol. **19**(11), 951–962 (2020)
9. Choi, H., Ha, S., Kang, H., Lee, H., Lee, D.S.: Deep learning only by normal brain PET identify unheralded brain anomalies. EBioMedicine **43**, 447–453 (2019)
10. Fernando, T., Gammulle, H., Denman, S., Sridharan, S., Fookes, C.: Deep learning for medical anomaly detection - a survey. ACM Comput. Surv. **54**(7), 1–37 (2021)

11. Ghosh, P., Sajjadi, M.S., Vergari, A., Black, M., Schölkopf, B.: From variational to deterministic autoencoders (2019). arXiv:1903.12436
12. Goodfellow, I., et al.: Generative adversarial nets. In: Advances in NeurIPS, vol. 27 (2014)
13. Hassanaly, R., Bottani, S., Sauty, B., Colliot, O., Burgos, N.: Simulation-based evaluation framework for deep learning unsupervised anomaly detection on brain FDG PET. In: SPIE Medical Imaging (2023)
14. Higgins, I., et al.: beta-VAE: learning basic visual concepts with a constrained variational framework. In: ICLR (2017)
15. Ho, J., Jain, A., Abbeel, P.: Denoising diffusion probabilistic models. In: Advances in NeurIPS, vol. 33, pp. 6840–6851 (2020)
16. Jack, C.R., et al.: A/T/N: an unbiased descriptive classification scheme for Alzheimer disease biomarkers. Neurology 87(5), 539–547 (2016)
17. Jagust, W.J., et al.: The Alzheimer's disease neuroimaging initiative positron emission tomography core. Alzheimer's Dement. 6(3), 221–229 (2010)
18. Kim, H., Mnih, A.: Disentangling by factorising. In: ICML, pp. 2649–2658. PMLR (2018)
19. Kingma, D.P., Welling, M.: Auto-encoding variational Bayes. In: ICLR 2014 - arXiv:1312.6114 (2014)
20. Kingma, D.P., Welling, M.: An Introduction to Variational Autoencoders. Now publishers Inc, Norwell (2019)
21. Kingma, D.P., Salimans, T., Jozefowicz, R., Chen, X., Sutskever, I., Welling, M.: Improved variational inference with inverse autoregressive flow. In: Advances in NeurIPS, vol. 29 (2016)
22. Larsen, A.B.L., Sønderby, S.K., Larochelle, H., Winther, O.: Autoencoding beyond pixels using a learned similarity metric. In: ICML, pp. 1558–1566. PMLR (2016)
23. Makhzani, A., Shlens, J., Jaitly, N., Goodfellow, I., Frey, B.: Adversarial autoencoders (2015). arXiv:1511.05644
24. Mostapha, M., et al.: Semi-supervised VAE-GAN for out-of-sample detection applied to MRI quality control. In: Shen, D., et al. (eds.) Medical Image Computing and Computer Assisted Intervention – MICCAI 2019. Lecture Notes in Computer Science(), vol. 11766, pp. 127–136. Springer, Cham (2019). https://doi.org/10.1007/978-3-030-32248-9_15
25. Rezende, D., Mohamed, S.: Variational inference with normalizing flows. In: ICML, pp. 1530–1538. PMLR (2015)
26. Routier, A., et al.: Clinica: an open-source software platform for reproducible clinical neuroscience studies. Front. Neuroinform. 15, 689675 (2021)
27. Snell, J., Ridgeway, K., Liao, R., Roads, B.D., Mozer, M.C., Zemel, R.S.: Learning to generate images with perceptual similarity metrics. In: ICIP, pp. 4277–4281. IEEE (2017)
28. Thibeau-Sutre, E., et al.: ClinicaDL: an open-source deep learning software for reproducible neuroimaging processing. Comput. Meth. Prog. Bio. 220, 106818 (2022)
29. Tolstikhin, I., Bousquet, O., Gelly, S., Schoelkopf, B.: Wasserstein auto-encoders. In: ICLR (2018)
30. Tomczak, J., Welling, M.: VAE with a VampPrior. In: International Conference on Artificial Intelligence and Statistics, pp. 1214–1223. PMLR (2018)
31. Uzunova, H., Schultz, S., Handels, H., Ehrhardt, J.: Unsupervised pathology detection in medical images using conditional variational autoencoders. IJCARS 14, 451–461 (2019)

32. Van Den Oord, A., et al.: Neural discrete representation learning. In: Advances in NeurIPS, vol. 30 (2017)
33. Wang, Z., Bovik, A.C., Sheikh, H.R., Simoncelli, E.P.: Image quality assessment: from error visibility to structural similarity. IEEE Trans. Image Process. **13**(4), 600–612 (2004)
34. Wen, J., et al.: Convolutional neural networks for classification of Alzheimer's disease: overview and reproducible evaluation. Media **63**, 101694 (2020)
35. Xia, T., Chartsias, A., Tsaftaris, S.A.: Pseudo-healthy synthesis with pathology disentanglement and adversarial learning. Media **64**, 101719 (2020)
36. Zhao, S., Song, J., Ermon, S.: InfoVAE: balancing learning and inference in variational autoencoders. In: Proceedings AAAI Conference on Artificial Intelligence, vol. 33, pp. 5885–5892 (2019)

Characterizing the Features of Mitotic Figures Using a Conditional Diffusion Probabilistic Model

Cagla Deniz Bahadir[1]([✉])[iD], Benjamin Liechty[2][iD], David J. Pisapia[2][iD], and Mert R. Sabuncu[3,4][iD]

[1] Cornell University and Cornell Tech, Biomedical Engineering, New York, NY, USA
cdb232@cornell.edu
[2] Weill Cornell Medicine, Pathology and Laboratory Medicine, New York, NY, USA
[3] Weill Cornell Medicine, Radiology, New York, NY, USA
[4] Cornell University and Cornell Tech, Electrical and Computer Engineering, New York, NY, USA

Abstract. Mitotic figure detection in histology images is a hard-to-define, yet clinically significant task, where labels are generated with pathologist interpretations and where there is no "gold-standard" independent ground-truth. However, it is well-established that these interpretation based labels are often unreliable, in part, due to differences in expertise levels and human subjectivity. In this paper, our goal is to shed light on the inherent uncertainty of mitosis labels and characterize the mitotic figure classification task in a human interpretable manner. We train a probabilistic diffusion model to synthesize patches of cell nuclei for a given mitosis label condition. Using this model, we can then generate a sequence of synthetic images that correspond to the same nucleus transitioning into the mitotic state. This allows us to identify different image features associated with mitosis, such as cytoplasm granularity, nuclear density, nuclear irregularity and high contrast between the nucleus and the cell body. Our approach offers a new tool for pathologists to interpret and communicate the features driving the decision to recognize a mitotic figure.

Keywords: Mitotic Figure Detection · Conditional Diffusion Models

1 Introduction

Mitotic figure (MF) count is an important diagnostic parameter in grading cancer types including meningiomas [9], breast cancer [18,20], uterine cancer [24], based on criteria set by World Health Organization (WHO) [9,18,24]. Due to the size of the whole slide images (WSIs), scarcity of MFs, and that presence of cells that mimic the morphological features of MFs, the detection process by pathologists is time consuming, subjective, and prone to error. This has led to the development of algorithmic methods that try to automate this process [3,6,14,16,23].

Earlier studies have used classical machine learning models like SVMs or random forests [2,5] to automate MF detection. Some of these works relied on morphological, textural and intensity-based features [14,17]. More recently, modern

A. Mukhopadhyay et al. (Eds.): DGM4MICCAI 2023, LNCS 14533, pp. 121–131, 2024.
https://doi.org/10.1007/978-3-031-53767-7_12

deep learning architectures for object detection, such as RetinaNet and Mask-RCNN have been employed for candidate selection, which is in turn followed by a ResNet or DenseNet-style classification model for final classification [8,22]. Publicly available datasets and challenges such as Canine Cutaneous Mast Cell Tumor (CCMCT) [6] and MIDOG 2022 [4], have catalyzed research in this area.

An important aspect of MF detection is that there is no independent ground-truth, other than human-generated labels. However, high inter-observer variability of human annotations has been documented [11]. In a related paper [7], the examples in the publicly available TUPAC16 dataset were re-labeled by two pathologists with the aid of an algorithm, which revealed that the updated labels can markedly change the resultant F1-scores. In another study, pathologists found the lack of being able to change the z axis focus, a limiting factor for determining MFs in digitized images [21]. Despite high inter-observer variability and inherent difficulty of recognizing MFs, all prior algorithmic work in this area relies on discrete labels, e.g. obtained via a consensus [1,4,6]. One major area that hasn't been studied is the morphological features of the cell images that contribute to the uncertainty in labeling MF.

Diffusion probabilistic models (DPMs) have recently been used to generate realistic images with or without conditioning on a class [10], including in biomedical applications [12,15]. In this paper, we trained a conditional DPM to synthesize cell nucleus images corresponding to a given mitosis score (which can be probabilistic). The synthetic images were in turn input to a pre-trained MF classification model to validate that the generated images corresponded to the conditioned mitosis score. The prevailing MF classification literature considers MF labels as binary. Embracing a probabilistic approach with the help of DPMs offers the opportunity to characterize the MF features, thus improves the interpretability of the MF classification process. Our DPM allows us to generate synthetic cell images that illustrate the transition from a non-mitotic figure to a definite mitosis. We also present a novel approach to transform a real non-mitotic cell into a mitotic version of itself, using the DPM. Overall, these analyses allow us to reveal and specify the image features that drive the interpretation of expert annotators and explain the sources of uncertainty in the labels.

2   Methodology

2.1   Datasets

Canine Cutaneous Mast Cell Tumor: CCMCT is a publicly available dataset that comprises 32 whole-slide images (WSI) [6]. We used the ODAEL (Object Detection Augmented Expert Labeled) variant which had the highest number of annotations. The dataset has a total of 262,481 annotated nuclei (based on the consensus of two pathologists). 44,880 of them are mitotic figures and the remainder are negative examples. We used the same training and test split as the original paper: 21 slides for training and 11 slides for test.

Meningioma: We created a meningioma MF dataset using 7 WSIs from a public dataset (The Digital Brain Tumour Atlas [13,19]) and 5 WSIs from our

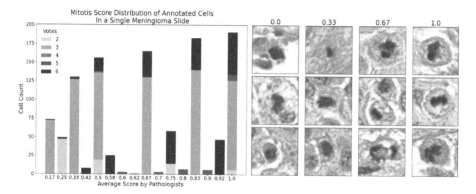

Fig. 1. Label distribution from a meningioma slide (left) and real cell images from a meningioma slide labeled with pathologists' scoring (right).

own hospital. The WSIs were non-exhaustively annotated for mitotic figures by two pathologists. The pathologists were allowed to give a score of 0, 0.5 and 1 to each candidate cell, 0.5 representing possible MFs where the annotator was not fully certain. There are a total of 6945 annotations, with 4186 negatives, 1439 definite MFs and 1320 annotations with scores varying between 0 and 1.

For 11 slides in the dataset each cell was seen by one of our two pathologists. For one slide in the dataset a total of 2095 cells were annotated by two pathologists using an in-house software that showed the cells to the annotators repetitively in a shuffled manner. The 2095 cells were independently annotated by both pathologists, up to three times each, yielding 6138 annotation instances. Figure 1 (left) shows the distribution of scores given to cells that are possibly MFs (i.e. have a non-zero score). The bar plot has also been color coded to show how many votes have been gathered for each score. This figure clearly demonstrates the probabilistic nature of the ground-truth label. Figure 1-right shows 12 cell examples with 4 different scores. Each of these cells has been annotated at least 3 times. The uncertainty of the scoring can be explained by visual inspection. In column 0, the pathologists are in consensus that these examples are not mitotic. In the top two examples, cytoplasmic membranes are indistinct or absent and the basophilic material that might prompt consideration of a mitotic figure does not have the shape or cellular context of condensed chromatin that would be encountered in a MF. The bottom example is a cell with a hyperchromatic nucleus that is not undergoing mitosis. In the examples with a 0.33 score, we observe that there is some density in the chromatin which can confound the interpreter. Yet, the lack of cellular borders make it challenging to interpret the basophilic material, and the putative chromatin appears smoother and lighter compared to the more confident examples. The second row in 0.67 is likely an example of telophase, and the uncertainty is attributable to the lighter and blurry appearance of the upper daughter cell. Also the potential anaphase in the third row of 0.67 shows an indistinct to absent cytoplasmic membrane, introducing uncertainty for a definitive mitotic figure call. The cell examples in column 1.0 depict strong features such as high chromatin density and irregularity in the nucleus and are good examples of metaphase due to the elongated appearance.

2.2 Training

We trained a ResNet34 model on the CCMCT dataset training set (21 WSIs) to discriminate mitotic and non-mitotic cells. The model was initialized with ImageNet pretrained weights. We used the Adam optimizer with a 1e−05 learning rate, batch size of 128, and binary cross entropy loss. The mitotic figures were sampled randomly in each mini batch with an equal number of randomly sampled negative examples. 64×64 RGB tiles were created around each cell nucleus and re-sampled to 256×256 to match Resnet34's input requirement. Each WSI was used in training and validation, with a 75% division in the y axis. The model was trained until validation loss converged. We randomly initialized with 3 different seeds during training and ensembled the three converged models by averaging to obtain final scores. This ResNet34 ensemble reached 0.90 accuracy and 0.81 F1-score in test data, similar to the numbers reported in the CCMCT paper with ResNet18 [6].

For the conditional DPM, we used an open-source PyTorch implementation[1]. The input embedding layer was changed into a fully connected layer that accepts a real-valued (probabilistic) input. Two DPMs were trained. One model was trained on the CCMCT dataset with the 11 test slides that were not seen by the classification model during training. Importantly, the CCMCT includes binary (i.e., deterministic) labels. The second DPM was trained on the meningioma dataset that comprised 12 slides with probabilistic labels. The models were trained on 64×64 images, with Adam optimizer, learning rate of 1e-04 and a batch size of 128. The trained model weights, code, and our expert annotations for The Brain Tumor Atlas Dataset is made publicly available[2].

2.3 Inference

During inference time, for a given random noise seed input, the DPM was run with a range of scores between 0 and 1, at 0.1 increments, as the conditional input value. This way, we generated a sequence of synthetic cell images that corresponds to a random generated cell nucleus transitioning into the mitosis stage.

We were also interested in using the DPMs to visualize the transformation of a *real* non-mitotic cell nucleus into a MF. To achieve this, we ran the DPM in forward diffusion mode, starting from a real non-mitotic cell image and iteratively adding noise. At intermediate time-points, we would then stop and invert the process to run the DPM to denoise the image - this time, conditioning on an MF score of 1 (i.e., definite mitosis). This allowed us to generate a sequence of

---

[1] https://github.com/lucidrains/denoising-diffusion-pytorch.
[2] https://github.com/cagladbahadir/dpm-for-mitotic-figures.

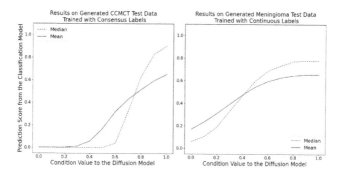

Fig. 2. Generated images from the diffusion model is tested on the ResNet34 classification model. The plot on the left shows the ResNet34 scores on the CCMCT generated test data and the plot on the right shows the scores for the synthetic meninigoma data. The shaded regions correspond to the standard error of the distribution of classification scores for every condition value.

images, each corresponding to different stopping time-points. Note that earlier stopping time-points yield images that look very similar to the original input non-mitotic image. However, beyond a certain time-point threshold, the synthesized image looks like a mitotic version of the input. We can interpret this threshold as a measure of how much an input image resembles a mitotic figure.

3   Results

3.1   Generated Cells with Probabilistic Labels

1000 sets of cells were generated with the two DPMs. Each set is a series of 11 synthetic images generated with the same random seed and condition values between 0 and 1, with 0.1 increments. The generated images were then input to the ResNet34 classification model. Figure 2 visualizes the average ResNet34 model scores for synthetic images generated for different condition values.

The CCMCT results presented on the left shows a steep increase in median classification scores given by the prediction model, starting from the conditional value of 0.6. The meningioma model which was trained on probabilistic labels, shows a more gradual trend of increase in the classification scores. This difference is likely due to fact that the CCMCT DPM was trained with binary labels and thus was not exposed to probabilistic conditional values, whereas the meningioma DPM was trained with probabilistic scores. The utilization of continuous labels obtained from pathologist votes during the training of DPMs enhances the comprehension of underlying factors that contribute to labeling uncertainty.

3.2   Selecting Good Examples

In order to select good synthetic images to interpret, we passed each set to the classification model. The identification of "good synthetic" images is contingent upon task definition and threshold selection. In our study, we adopted

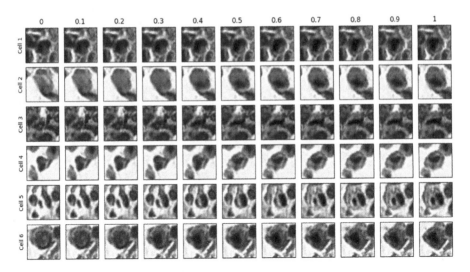

Fig. 3. Synthetic cell sets generated with the CCMT DPM. Rows represent cell sets generated from the same random noise, transitioning from non-mitotic (left) to definite mitotic figures (right). Condition values are listed above.

a selection process focused on visually evaluating smooth transitions between synthetic images. If a synthetic image set started with predicted mitosis score of less than 0.1, reached a final score greater than 0.9 and the sequence of scores were relatively smooth (e.g., the change in scores was less than 0.30 between each increment), we included it in our visual analysis presented below.

3.3   CCMCT Visual Results

Figure 3 shows the selected examples from the diffusion model trained on the CCMCT dataset. Each row represents one set that was generated with the same random seed. This visual depiction shows that there are a wide variety of morphological changes that correlate with transitioning from a non-mitotic to a mitotic cell. In cells 1 and 5 there are several dark areas at lower probabilities in the first image. The model merges several of the spots to one, bigger and darker chromosomal aggregate in cell 1 and two dense chromosomal aggregates in cell 5, mimicking telophase. The surrounding dark spots are faded away and the cell membrane becomes more defined. In cells 2 and 6, condensed chromatin is generated from scratch as the conditional value is increased. In cell 3, the faint elongated structure in the center of the image gradually merges into a denser, more defined, irregular and elongated chromosomal aggregate, mimicking metaphase. In cell 4, the cell body size increases around the nucleus while gradually creating more contrast by lowering the darkness of the cytoplasm, sharpening the edges and increasing the density of the chromatin.

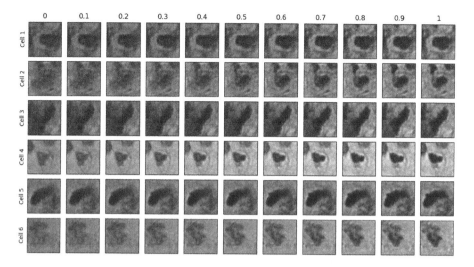

Fig. 4. Sets of synthetic cells generated with the Meningioma DPM. Each row illustrates a non-mitotic nucleus transitioning to a mitotic figure.

3.4   Meningioma Visual Results

Figure 4 shows selected examples from the DPM trained on the meningioma dataset. In cells 1, 4 and 5, the nucleus size stays roughly the same while the chromatin gets darker and sharper, accentuating the irregularity in the chromatin edges. In cells 2 and 6, the dense chromatin is created from a blurry and indistinct starting image. The sharp, dense and granular chromatin in cell 6 eventually resembles prophase. In cell 3 the intensity of the chromatin increases while elongating and sharpening, resembling metaphase. Note that, the differences between the first and last columns are in general very subtle to the untrained eye.

3.5   Generating Mitotic Figures from Real Negative Examples

Mitotic figure variants can also be generated from real negative examples. In a DPM, the forward diffusion mode gradually adds noise to an input image over multiple time-points. Inverting this process, allows us to convert a noisy image into a realistic looking image. When running the inverse process, we conditioned on a mitosis score of 1, ensuring that the DPM attempted to generate a mitotic figure. Note that, in our experiments below, we always input a non-mitotic image and run the forward and backward modes for a varying number of time-points. The longer we run these modes, the less the output image resembles the input and the more mitotic it looks.

In Fig. 5, 6 negative cell examples from 6 different WSIs are shown in the real column. Each column corresponds to a different number of time-points that we used to run in the forward and inverse modes. We can appreciate that the real negative cell examples are slowly transitioning into MFs, with a variety of changes occurring in intermediate steps. Each cell was annotated by a pathologist retrospectively to mark the earliest time-point where the cell resembles a MF, indicated with the yellow frame, and when the example becomes a convincing MF, indicated with the green frame. We can see that the examples that reach the yellow frame earlier, such as cells 1 and 6, already exhibit features associated with mitosis, such as dense and elongated chromatin. Conversely, examples that reach the yellow frame later can be thought of as more obvious negative examples. Examples such as cell 3, where the jump from the yellow to green state occurs in a single frame, allow us to isolate the changes needed for certainty. In this case, we observe that the increased definition of the cell body while the chromatin remains the same, is the driving feature. Overall, the relative time points in which the real negative cell examples morph into a mitotic figure can also be used as a marker to determine which examples are likely to be mistaken for mitotic figures by classification models, and the data sets can be artificially enhanced for those examples to reduce misclassification.

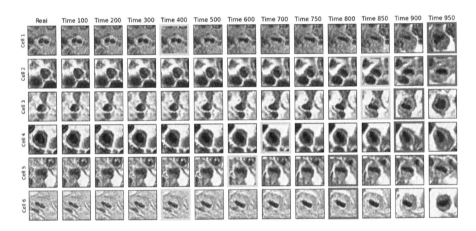

Fig. 5. Generated Mitotic Figures from real negative examples from CCMCT dataset during intermediate time points.

4   Discussion

Detecting mitotic figures is a clinically significant, yet difficult and subjective task. There is no gold-standard ground-truth and we rely on expert pathologists for annotations. However, the variation and subjectivity of pathologist annotations is well documented. To date, there has been little focus on the uncertainty in pathologist labels and no systematic effort to specify the image features that drive a mitosis call.

In this paper, we proposed the use of a diffusion probabilistic model, to characterize the MF detection task. We presented strategies for visualizing how a (real or synthetic) non-mitotic nucleus can transition into a mitosis state. We observed that there is a wide variety of features associated with this transition, such as intensity changes, sharpening, increase of irregularity, erasure of certain spots, merging of features, redefinition of the cell membrane and increased granularity of the chromatin or cytoplasm.

There are several directions we would like to explore in the future. The conditional DPM can be used to enrich the training data or generate training datasets comprising only of synthetic images to train more accurate MF detection models, particularly in limited sample size scenarios. The synthetic images can be used to achieve more useful ground-truth labels that do not force pathologists to an arbitrary consensus, but instead allows them to weigh the different features in order to converge on a probabilistic annotation. These probabilistic labels can, in turn, yield more accurate, calibrated and/or robust nucleus classification tools. We also are keen to extend the DPM to condition on different nucleus types and tissue classes.

Acknowledgements. Funding for this project was partially provided by The New York-Presbyterian Hospital William Rhodes Center for Glioblastoma-Collaborative Research Initiative, a Weill Cornell Medicine Neurosurgery-Cornell Biomedical Engineering seed grant, The Burroughs Wellcome Weill Cornell Physician Scientist Program Award, NIH grant R01AG053949, and the NSF CAREER 1748377 grant. Project support for this study was provided by the Center for Translational Pathology of the Department of Pathology and Laboratory Medicine at Weill Cornell Medicine.

References

1. Albarqouni, S., Baur, C., Achilles, F., Belagiannis, V., Demirci, S., Navab, N.: Aggnet: deep learning from crowds for mitosis detection in breast cancer histology images. IEEE Trans. Med. Imaging **35**(5), 1313–1321 (2016)
2. Albayrak, A., Bilgin, G.: Mitosis detection using convolutional neural network based features. In: 2016 IEEE 17th International symposium on computational intelligence and informatics (CINTI), pp. 000335–000340. IEEE (2016)
3. Aubreville, M., et al.: Deep learning algorithms out-perform veterinary pathologists in detecting the mitotically most active tumor region. Sci. Rep. **10**(1), 1–11 (2020)
4. Aubreville, M., et al.: Mitosis domain generalization in histopathology images-the midog challenge. Med. Image Anal. **84**, 102699 (2023)
5. Beevi, K.S., Nair, M.S., Bindu, G.: Automatic mitosis detection in breast histopathology images using convolutional neural network based deep transfer learning. Biocybern. Biomed. Eng. **39**(1), 214–223 (2019)
6. Bertram, C.A., Aubreville, M., Marzahl, C., Maier, A., Klopfleisch, R.: A large-scale dataset for mitotic figure assessment on whole slide images of canine cutaneous mast cell tumor. Sci. Data **6**(1), 1–9 (2019)

7. Bertram, C.A., et al.: Are pathologist-defined labels reproducible? Comparison of the TUPAC16 mitotic figure dataset with an alternative set of labels. In: Cardoso, J., et al. (eds.) IMIMIC/MIL3ID/LABELS -2020. LNCS, vol. 12446, pp. 204–213. Springer, Cham (2020). https://doi.org/10.1007/978-3-030-61166-8_22

8. Fick, R.H.J., Moshayedi, A., Roy, G., Dedieu, J., Petit, S., Hadj, S.B.: Domain-specific cycle-GAN augmentation improves domain generalizability for mitosis detection. In: Aubreville, M., Zimmerer, D., Heinrich, M. (eds.) MICCAI 2021. LNCS, vol. 13166, pp. 40–47. Springer, Cham (2022). https://doi.org/10.1007/978-3-030-97281-3_5

9. Ganz, J., et al.: Automatic and explainable grading of meningiomas from histopathology images. In: MICCAI Workshop on Computational Pathology, pp. 69–80. PMLR (2021)

10. Ho, J., Jain, A., Abbeel, P.: Denoising diffusion probabilistic models. In: Advances in Neural Information Processing Systems, vol. 33, pp. 6840–6851 (2020)

11. Malon, C., et al.: Mitotic figure recognition: agreement among pathologists and computerized detector. Anal. Cell. Pathol. 35(2), 97–100 (2012)

12. Moghadam, P.A., et al.: A morphology focused diffusion probabilistic model for synthesis of histopathology images. In: Proceedings of the IEEE/CVF Winter Conference on Applications of Computer Vision, pp. 2000–2009 (2023)

13. Roetzer-Pejrimovsky, T., et al.: the digital brain Tumour atlas, an open histopathology resource. Sci. Data 9(1), 55 (2022)

14. Saha, M., Chakraborty, C., Racoceanu, D.: Efficient deep learning model for mitosis detection using breast histopathology images. Comput. Med. Imaging Graph. 64, 29–40 (2018)

15. Sanchez, P., Kascenas, A., Liu, X., O'Neil, A.Q., Tsaftaris, S.A.: What is healthy? Generative counterfactual diffusion for lesion localization. In: Mukhopadhyay, A., Oksuz, I., Engelhardt, S., Zhu, D., Yuan, Y. (eds.) DGM4MICCAI 2022. LNCS, vol. 13609, pp. 34–44. Springer, Cham (2022). https://doi.org/10.1007/978-3-031-18576-2_4

16. Sebai, M., Wang, T., Al-Fadhli, S.A.: Partmitosis: a partially supervised deep learning framework for mitosis detection in breast cancer histopathology images. IEEE Access 8, 45133–45147 (2020)

17. Sigirci, I.O., Albayrak, A., Bilgin, G.: Detection of mitotic cells in breast cancer histopathological images using deep versus handcrafted features. Multimedia Tools Appl. 81(10), 13179–13202 (2022)

18. Sohail, A., Khan, A., Wahab, N., Zameer, A., Khan, S.: A multi-phase deep CNN based mitosis detection framework for breast cancer histopathological images. Sci. Rep. 11(1), 1–18 (2021)

19. Thomas, R.P., et al.: The digital brain Tumour atlas, an open histopathology resource [data set]. https://doi.org/10.25493/WQ48-ZGX

20. Veta, M., van Diest, P.J., Pluim, J.P.: Detecting mitotic figures in breast cancer histopathology images. In: Medical Imaging 2013: Digital Pathology, vol. 8676, pp. 70–76. SPIE (2013)

21. Wei, B.R., et al.: Agreement in histological assessment of mitotic activity between microscopy and digital whole slide images informs conversion for clinical diagnosis. Acad. Pathol. 6, 2374289519859841 (2019)

22. Wilm, F., Marzahl, C., Breininger, K., Aubreville, M.: Domain adversarial RetinaNet as a reference algorithm for the MItosis DOmain generalization challenge. In: Aubreville, M., Zimmerer, D., Heinrich, M. (eds.) MICCAI 2021. LNCS, vol. 13166, pp. 5–13. Springer, Cham (2022). https://doi.org/10.1007/978-3-030-97281-3_1

23. Wu, B., et al.: FF-CNN: an efficient deep neural network for mitosis detection in breast cancer histological images. In: Valdés Hernández, M., González-Castro, V. (eds.) MIUA 2017. CCIS, vol. 723, pp. 249–260. Springer, Cham (2017). https://doi.org/10.1007/978-3-319-60964-5_22

24. Zehra, T., et al.: A novel deep learning-based mitosis recognition approach and dataset for uterine leiomyosarcoma histopathology. Cancers **14**(15), 3785 (2022)

A 3D Generative Model of Pathological Multi-modal MR Images and Segmentations

Virginia Fernandez$^{(\boxtimes)}$ ⓘ, Walter Hugo Lopez Pinaya ⓘ, Pedro Borges ⓘ,
Mark S. Graham ⓘ, Tom Vercauteren ⓘ, and M. Jorge Cardoso ⓘ

King's College London, London WC2R 2LS, UK
virginia.fernandez@kcl.ac.uk

Abstract. Generative modelling and synthetic data can be a surrogate for real medical imaging datasets, whose scarcity and difficulty to share can be a nuisance when delivering accurate deep learning models for healthcare applications. In recent years, there has been an increased interest in using these models for data augmentation and synthetic data sharing, using architectures such as generative adversarial networks (GANs) or diffusion models (DMs). Nonetheless, the application of synthetic data to tasks such as 3D magnetic resonance imaging (MRI) segmentation remains limited due to the lack of labels associated with the generated images. Moreover, many of the proposed generative MRI models lack the ability to generate arbitrary modalities due to the absence of explicit contrast conditioning. These limitations prevent the user from adjusting the contrast and content of the images and obtaining more generalisable data for training task-specific models. In this work, we propose brainSPADE3D, a 3D generative model for brain MRI and associated segmentations, where the user can condition on specific pathological phenotypes and contrasts. The proposed joint imaging-segmentation generative model is shown to generate high-fidelity synthetic images and associated segmentations, with the ability to combine pathologies. We demonstrate how the model can alleviate issues with segmentation model performance when unexpected pathologies are present in the data.

1 Introduction

In the past decade, it has been shown that deep learning (DL) has the potential to ease the work of clinicians in tasks such as imaging segmentation [10], an otherwise time-consuming task that requires expertise in the imaging modality and anatomy. Nonetheless, the performance and generalisability of DL algorithms is linked to how extensive and unbiased the training dataset is [14]. While large image datasets in computer vision are widely available [9], this is not the case

Supplementary Information The online version contains supplementary material available at https://doi.org/10.1007/978-3-031-53767-7_13.

A. Mukhopadhyay et al. (Eds.): DGM4MICCAI 2023, LNCS 14533, pp. 132–142, 2024.
https://doi.org/10.1007/978-3-031-53767-7_13

for medical imaging because images are harder to acquire and share, as they are subject to tight data regulations [23]. In addition, most state-of-the-art segmentation algorithms are supervised and require labels as well as images, and because obtaining these requires substantial time and expertise, they tend to focus on a specific region of interest, making dataset harmonisation harder. In brain MRI, where studies are tailored to a pathology and population of interest, obtaining a large, annotated, multi-modal and multi-pathological dataset is challenging. An option to overcome this is to resort to domain randomisation methods such as SynthSeg [5], but their performance in the presence of highly variable pathologies such as tumours has not been tackled. Alternatively, data augmentation via deep generative modelling, an unsupervised DL branch that learns the input data distribution, has been applied in recent years to enrich existing medical datasets, producing realistic, usable synthetic data with the potential to complement [3] or even replace [11] real datasets, using architectures such as generative adversarial networks (GANs) and the more recent diffusion models (DMs) [21,26]. One of the major roadblocks, though, when it comes to applying synthetic data to segmentation tasks is that of producing labelled data. Conditioning can give the user some control over the generated phenotypes, such as age [21]. However, to our knowledge, only a handful of works deliver segmentations to accompany the data. In the case of published models generating data based on real labels [22], we must consider that labels are not usually shared, as they are sometimes considered protected health information, due to the risk of patient re-identification [27], and due to the above-mentioned difficulty to produce. Therefore, it may be beneficial that the labels themselves are also algorithmically generated from a stochastic process. Few works in the literature provide synthetic segmentations [4,11,12], especially enclosing healthy and multiple diseased regions. Among the latter, these models are limited to 2D, and they hardly allow the user to modulate their content (e.g., selecting the subject's pathology or age), limiting their applicability.

Contributions: in this work, we propose a 3D generative model of the brain that provides multi-modal brain MR images and corresponding semantic maps generated by giving the user the power to condition on the pathological phenotype of the synthetic subject. We showcase the benefits of using these synthetic datasets on a downstream white matter hyperintensity (WMH) segmentation task when the test dataset contains also contains tumours.

2   Methods

2.1   Data

For training, we used the SABREv3 dataset consisting of 630 T1, FLAIR and T2 images [17], a subset of 66 T1 and FLAIR volumes from ADNI2 [16], and 103 T1, FLAIR and T2 volumes from a set of sites from BRATS [1,2,19]. Due to the large computational costs associated with training generative models, it is not tenable to train them using full resolution, full size 3D images: we circumvented this issue by, on one hand, mapping images to a 2 mm isotropic MNI space, resulting in volumes of dimensions $96 \times 128 \times 96$, and on the other hand, by operating with patches, taken from 1 mm data of dimensions $146 \times 176 \times 112$. Bronze-standard partial volume (PV) maps of the cerebrospinal fluid (CSF), grey matter (GM), white matter (WM), deep grey matter (DGM) and brainstem were obtained using GIF [6], masking out tumours for BRATS. These healthy labels were overlaid with manual lesion labels provided with the datasets: WMH for the first two; and gadolinium-enhancing (GDE), non-enhancing (nGDE) tumour and edema for BRATS.

2.2   Algorithm

Our pipeline consists of a conditional generator of semantic maps and an image generator, depicted in Fig. 1. It is based on the generative model proposed in [11]: a label generator, consisting of a latent diffusion model (LDM), is trained on the healthy tissue and lesion segmentations. Independently, a SPADE-like [20] network is trained on the PV maps and the multi-modal images. For the 1 mm³ data, patches of $146 \times 176 \times 64$ had to be used for the image generator.

Label Generator: The proposed label generator is based on a latent diffusion model (LDM), made up of a spatial variational auto-encoder (VAE) and a diffusion model (DM) operating in its latent space. The VAE of the 2 mm³ resolution model had 3 downsamplings, and the 1 mm³ had 4, resulting in latent spaces of shapes $32 \times 24 \times 32$ and $24 \times 24 \times 16$, respectively. The VAE is trained using focal loss ($\gamma = 3$), Kullback-Leibler distance (KLD) loss to stabilise the latent space, Patch-GAN adversarial loss and a perceptual loss based on the features of MED3D [7], implemented using MONAI [8]. For the diffusion model, we predict the velocity using the *v-parametrization* approach from [24] and optimise it

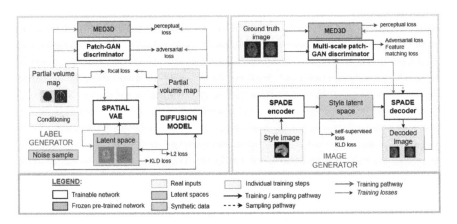

Fig. 1. Architecture of our two-stage model: the left block corresponds to the label generator, and the right block to the image generator. Training and inference pathways are differentiated with black, red and dashed arrows. (Color figure online)

via an l_2 loss. We used T = 1000 timesteps. We used a PNDM [18] scheduler to sample data, predicting only 150 timesteps. In addition, disease conditioning dc was applied using a cross-attention mechanism. For each subject j and disease type l, we have a label map M_{jl}, from which we produce a conditioning value dc_{jl} reflecting the voxels labelled as l in the map, normalised by the maximum number of l voxels across the dataset, i.e.:

$$dc_{jl} = \frac{\sum_{n=1}^{N} M_{jl}}{max_j \sum_{n=1}^{N} M_{jl}}, \tag{1}$$

where the sum is across all voxels in the map. We trained the VAE for 250 epochs and the DM for 400, on an NVIDIA A100 DGX node.

Image Generator: We modified the SPADE model used in [11] to extend the 2D generator and multi-scale discriminator to 3D. The encoder, which should only convert the contrast of the input image to a style vector, was kept as a 2D network, as it was found to work well while being parsimonious. To ensure that the most relevant brain regions informed the style, we used sagittal slices instead of axial ones as done in [11], selecting them randomly from the central 20 slices of each input volume. We kept the original losses from [20] to optimise the network. We replaced the network on which the perceptual loss is calculated with MED3D [7], as its features are also in 3D and fine-tuned on medical images, which are more domain-pertinent. We ran a full ablation study on the losses introduced in [11]; we dropped the modality-dataset discriminator loss, as it did not lead to major improvements but kept the slice consistency loss. To train the $1\,\text{mm}^3$ model, we used random patches of size 64 along the axial dimension. During inference, we used a sliding-window approach with a 5-slice overlap. We trained the networks for 350 epochs on an NVIDIA A100 GPU. Further details are provided in the supplementary materials. Code is available at https://github. com/virginiafdez/brainSPADE3D_rel.git.

2.3 Downstream Segmentation Task

To compare the performance in segmentation tasks of our synthetic datasets, we performed several experiments using nnUNetv2 [15] as a strong baseline architecture, adjusting only the number of epochs until convergence. The partial volume maps were converted to categorical labels via an *argmax* operator.

3 Experiments

3.1 Quality of the Generated Images and Labels Pairs

Without established baselines for paired 3D healthy and pathological labels and image pairs, we assess our synthetic data by comparing them to real data and showing how they can be applied to downstream segmentation tasks. Examples of generated labels and images are depicted in Fig. 2. In one case, we used a conditioning unseen by the model during training, *WMH + all tumour layers*: both $1\,\text{mm}^3$ and $2\,\text{mm}^3$ label generators show the capability of handling this unseen combination, resulting in both lesions being present in the resulting images. However, we observed a lower ability of extrapolating to unseen phenotypes in the $1\,\text{mm}^3$ model, with only about 37% of the labels inferred using such conditioning resulting in the desired phenotype being met, as opposed to the $2\,\text{mm}^3$ model which manifested both lesion types in 100% of the generated samples.

Quality of the Labels: As the label generator is stochastic, we cannot compute paired similarity metrics. Instead, we compare the number $V_{i,j}$, for image i and region j, of CSF, GM, WM, DGM and brainstem voxels across subjects between a synthetic dataset of 500 volumes and a subset of the training dataset of the same size, excluding tumour images, but allowing for low WMH values as these don't disrupt the anatomy of the brain. The number of voxels $V_{i,j}$ is calculated as: $V_{i,j} = \sum_{n=1}^{N}(i_{n==j})$, where N is the number of pixels in the image. Table 1 reports mean values and standard deviations, demonstrating that our model generates labels with mean volumes similar to real data. By comparing the labels, we saw that the considerable discrepancy between $V_{i,CSF}$ at $1\,\text{mm}^3$ is due to a loss of details in the subarachnoid CSF, which is very thin, likely due to the high number of VAE downsamplings at $1\,\text{mm}^3$ (visual comparison is available in

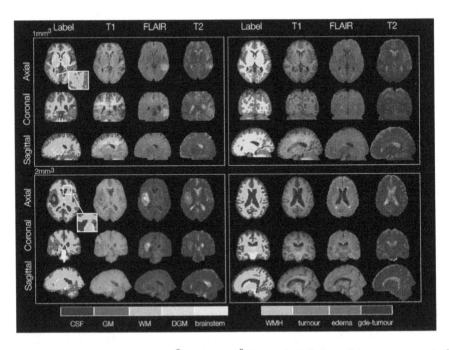

Fig. 2. Example synthetic $1\,\text{mm}^3$ and $2\,\text{mm}^3$ isotropic labels and images generated using tumour+WMH (left) and WMH (right) conditioning. The augmented frame in the top left images shows the small WMH lesions near the ventricles.

supplementary Fig. 1). However, we observe a lower standard deviation in the tissue volumes, indicating that the label generator does not capture the natural variability of brain tissues.

Table 1. Mean number of voxels and standard deviation of brain regions across real and synthetic datasets. Every value has been multiplied by 10^{-4}.

Dataset	CSF	GM	WM	DGM	Brainstem
Real ($1\,\mathrm{mm}^3$)	$20.342_{2.830}$	$37.634_{1.747}$	$44.872_{2.270}$	$4.301_{0.581}$	$1.238_{0.269}$
Synthetic ($1\,\mathrm{mm}^3$)	$14.141_{0.964}$	$43.583_{0.735}$	$42.029_{1.728}$	$6.713_{0.592}$	$1.066_{0.102}$
Real ($2\,\mathrm{mm}^3$)	$4.790_{0.573}$	$9.879_{0.585}$	$7.436_{0.512}$	$0.453_{0.080}$	$0.363_{0.036}$
Synthetic ($2\,\mathrm{mm}^3$)	$4.641_{0.163}$	$8.564_{0.297}$	$7.008_{0.198}$	$0.587_{0.978}$	$0.343_{0.013}$

Quality of the Generated Images: To assess the performance of our image generator, we use a hold-out test set of PV maps and corresponding T1, FLAIR and T2 images, to compute the structural similarity index (SSIM) between the ground truth images and the image generated when the real PV map and a slice from the ground truth were used as inputs to the models. The mean SSIMs obtained are summarised in Table 2, showing that the model performs similarly across different contrasts. Discrepancies between real and synthetic images can be explained by the stochasticity of the style encoder.

Synthetic Data for Healthy Region Segmentation: We assess the performance of our generated pairs on a CSF, GM, WM, DGM and brainstem segmentation task. We train an nnU-Net model, $M_{healthy}$ on the T1 volumes of the real subset of 500 subjects mentioned earlier, and $M'_{healthy}$ on the 500 synthetic T1 and label pairs, then test both models on a hold-out test set of 30 subjects from the SABRE dataset. The Dice scores on all regions are reported in Table 3. Although $M_{healthy}$ performs better in all regions, the $M'_{healthy}$ trained on purely synthetic data demonstrated a competitive performance for all regions except the DGM. DGM is a complex anatomical region comprising several small structures with intensities ranging between those typical for GM and WM. Thus, the PV map value for a voxel in this region will split its probability between DGM, WM and GM rather than favouring just one class, which is problematic for nnU-Net, as it requires categorical inputs to train the model, resulting in noisy ground truth labels that cause a larger distribution shift for the region. Examples of these noisy training and test labels are showcased in supplementary Fig. 2.

3.2   WMH Segmentation in the Presence of Tumour Lesions

Aim: The main aim of this work is to show how synthetic data can increase the performance of segmentation models when training datasets are biased towards a specific phenotype. We focus on WMH segmentation. Our target dataset is a

Table 2. SSIM values obtained between real and synthetic images for T1, FLAIR and T2 contrasts for both models, generated using real PV maps.

1 mm^3			2 mm^3		
T1	FLAIR	T2	T1	FLAIR	T2
$0.842_{0.083}$	$0.798_{0.082}$	$0.794_{0.075}$	$0.922_{0.010}$	$0.910_{0.030}$	$0.909_{0.025}$

subset of 30 unseen volumes from BRATS (a different set of sites from those seen by the generative model) containing WMH lesions and tumours. We hypothesis that a WMH segmentation model trained on images that do not contain tumours will label these as WMH, as tumours and WMH share some intensity similarities in the FLAIR contrast typically used to segment WMH [25]. With the proposed generative model, we can generate synthetic data containing subjects with *both* tumours and WMH, which should make the training model robust to cases where both diseases are present, therefore reducing false positives.

Table 3. Mean Dice score and standard deviation obtained for the models trained on real and synthetic data. Asterisks denote significantly better performance.

Resolution	Model	CSF	GM	WM	DGM	Brainstem
1 mm^3	$M_{healthy}$	$0.957_{0.005}$*	$0.959_{0.003}$*	$0.971_{0.003}$*	$0.875_{0.015}$*	$0.958_{0.021}$*
1 mm^3	$M'_{healthy}$	$0.884_{0.014}$	$0.912_{0.009}$	$0.936_{0.005}$	$0.684_{0.034}$	$0.874_{0.036}$
2 mm^3	$M_{healthy}$	$0.947_{0.057}$*	$0.958_{0.046}$*	$0.968_{0.039}$*	$0.887_{0.065}$*	$0.962_{0.024}$*
2 mm^3	$M'_{healthy}$	$0.869_{0.057}$	$0.895_{0.052}$	$0.931_{0.047}$	$0.703_{0.100}$	$0.905_{0.025}$

We ran this experiment with 2 mm^3 isotropic data, as the phenotype conditioning worked better (see Sect. 3.1) in this model. From a stack of 500 real FLAIR volumes from the SABRE dataset, and a stack of 500 FLAIR synthetic volumes generated from synthetic labels conditioned on both tumours and WMH, we train several models $M_{R_{PR}S_{PS}}$, varying the % proportions PR and PS of real and synthetic data respectively. In addition, even if the premise of this work is that users do not have access to real data containing tumours, we train model $M_{R_{WMH}R_{tum}}$ on the real FLAIR volumes from the SABRE dataset, *and* the BRATS tumour volumes used to train the synthetic model, leaving the training labels empty, as no prior WMH segmentations are available for BRATS. We calculate the Dice score on WMH on a hold-out test set of 30 subjects from the SABRE dataset. As we do not have WMH ground truth labels for our test set from BRATS, but we have tumour labels, we compute the proportion of tumour pixels incorrectly labelled as WMH and note this metric FP_{tum}.

Table 4. Mean Dice score, precision and recall obtained on the SABRE dataset by all the WMH segmentation models we trained, and FP_{tum} ratio on the BRATS holdout dataset. Asterisks indicate statistical significance (SAB: SABRE dataset).

Model	Dice (SAB) ↑	precision (SAB) ↑	recall (SAB) ↑	FP_{tum} (BRATS) ↓
$M_{R_{100}S_0}$	$0.728_{0.281}$*	$0.761_{0.236}$	$0.751_{0.132}$*	$0.325_{0.232}$
$M_{R_{75}S_{25}}$	$0.713_{0.208}$	$0.745_{0.231}$	$0.743_{0.137}$	$0.902_{0.187}$
$M_{R_{50}S_{50}}$	$0.716_{0.211}$*	$0.742_{0.227}$	$0.754_{0.132}$*	$0.108_{0.209}$
$M_{R_{25}S_{75}}$	$0.722_{0.210}$	$0.755_{0.225}$	$0.722_{0.138}$	$0.079_{0.171}$
$M_{R_5S_{95}}$	$0.642_{0.210}$	$0.716_{0.243}$	$0.628_{0.146}$	$0.023_{0.078}$
$M_{R_0S_{100}}$	$0.362_{0.147}$	$0.726_{0.294}$	$0.263_{0.104}$	$0.001_{0.002}$*
$M_{R_{WMH}R_{tum}}$	$0.710_{0.233}$	$0.742_{0.250}$	$0.741_{0.131}$	$0.026_{0.140}$*

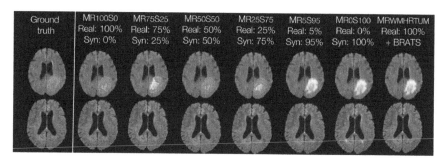

Fig. 3. Sample WMH predictions on the BRATS dataset (top) and the SABRE dataset (bottom) for all our models, in red. The leftmost column shows the tumour mask for the BRATS dataset (in blue) and the ground truth WMH for the SABRE dataset. (Color figure online)

Results: Results are reported in Table 4. Example segmentations and predicted WMH masks on both test sets are depicted in Fig. 3. $M_{R_{100}S_0}$ achieved the top Dice on WMH for its in-domain test set, but it has one of the worse FP_{tum} scores on the BRATS set. While $M_{R_0S_{100}}$ achieved a low Dice on the SABRE dataset, it had a FP_{tum} score that is significantly lower than that of $M_{tum-real}$. All the models trained on a combination of real and synthetic data achieve a competitive FP_{tum} without compromising the WMH Dice. While all models have a comparable precision, $M_{R_0S_{100}}$ has low recall; caused by an underestimation of WMH, as seen in Fig. 3. Interestingly, besides $M_{R_0S_{100}}$ and $M_{R_5S_{95}}$, the edematous area of the tumours still gets partially segmented as WMH. $M_{R_{WMH}R_{tum}}$ achieves very good Dice, precision, recall and FP_{tum} metrics; but, while examining the WMH segmentations on the BRATS dataset, all the segmentations were empty, as shown in Fig. 3, which indicates that, because the WMH training labels for BRATS were empty, the model has mapped the appearance and/or phenotype of the BRATS dataset to an absence of WMH.

4   Discussion and Conclusion

This work presents a label and multi-contrast brain MRI 3D image generator that can supplement real datasets in segmentation tasks for healthy tissues and pathologies. The synthetic data provided by our model can boost the precision and robustness of WMH segmentation models when tumours are present in the target dataset, showing the potential for having content and style-disentangled generative models that can combine the phenotypes seen in their training datasets. While disentanglement is covered in [11], 3D can help produce data usable in scenarios where the context of neighbouring slices is meaningful, such as segmenting small lesions. In addition, disease conditioning, which was not implemented in [11], can be challenging in 2D, as some diseases depend on the axial location, such as WMH. Our current set-up has, however, some limitations. First, there is a caveat in using $2\,mm^3$ isotropic data or patching at $1\,mm^3$. Even so, diffusion models for high resolution 3D images have to operate in a latent space that causes loss of semantic variability (See Sect. 3.1) and small details, affecting the downstream segmentation task, partly because a gap appears in the image synthesis process between synthetic and real labels. Further work should attempt to harmonise the semantic synthetic and real domains. Although one of the causes of this limitation is capacity, the latest advances in diffusion models show that higher performance and resolution can be achieved [13], potentially leading to better labels and, effectively, less domain shift between real and synthetic domains. Secondly, conditioning on variables such as age or ventricle size could also translate into more variability across the generated volumes [21], overcoming the limitation in tissue variability observed in Table 1. The method can be scaled to more pathologies and tasks, as model sharing allows for fine-tuning on more pathological labels, thus making segmentation models more generalisable to the diverse phenotypes of real brain MR data.

References

1. Bakas, S., et al.: Advancing the cancer genome atlas glioma MRI collections with expert segmentation labels and radiomic features. Sci. Data **4** (2017). https://doi.org/10.1038/SDATA.2017.117, https://pubmed.ncbi.nlm.nih.gov/28872634/
2. Bakas, S., et al.: Identifying the best machine learning algorithms for brain tumor segmentation, progression assessment, and overall survival prediction in the BRATS challenge. Sandra Gonzlez-Vill **124** (2018). https://arxiv.org/abs/1811.02629v3
3. Barile, B., Marzullo, A., Stamile, C., Durand-Dubief, F., Sappey-Marinier, D.: Data augmentation using generative adversarial neural networks on brain structural connectivity in multiple sclerosis. Comput. Methods Programs Biomed. **206** (2021). https://doi.org/10.1016/J.CMPB.2021.106113, https://pubmed.ncbi.nlm.nih.gov/34004501/
4. Basaran, B.D., Matthews, P.M., Bai, W.: New lesion segmentation for multiple sclerosis brain images with imaging and lesion-aware augmentation. Front. Neurosci. **16** (2022). https://doi.org/10.3389/FNINS.2022.1007453, https://pubmed.ncbi.nlm.nih.gov/36340756/

5. Billot, B., Magdamo, C., Arnold, S.E., Das, S., Iglesias, J.E.: Robust machine learning segmentation for large-scale analysis of heterogeneous clinical brain MRI datasets. Proc. Natl. Acad. Sci. **120**(9), e2216399120 (2022). https://doi.org/10.1073/PNAS.2216399120/SUPPL_FILE/PNAS.2216399120.SAPP.PDF, https://arxiv.org/abs/2209.02032
6. Cardoso, M.J., et al.: Geodesic information flows: spatially-variant graphs and their application to segmentation and fusion. IEEE Trans. Med. Imaging **34**(9), 1976–1988 (2015)
7. Chen, S., Ma, K., Zheng, Y.: MED3D: Transfer Learning for 3D Medical Image Analysis. https://github.com/Tencent/MedicalNet
8. Consortium, M.: MONAI: Medical Open Network for AI, March 2020
9. Deng, J., Dong, W., Socher, R., Li, L.J., Kai, L., Fei-Fei, L.: ImageNet: a large-scale hierarchical image database, pp. 248–255 (2010). https://doi.org/10.1109/CVPR.2009.5206848
10. Esteva, A., et al.: A guide to deep learning in healthcare. Nat. Med. **25**(1), 24–29 (2019). https://doi.org/10.1038/s41591-018-0316-z, https://www.nature.com/articles/s41591-018-0316-z
11. Fernandez, V., et al.: Can segmentation models be trained with fully synthetically generated data? In: Zhao, C., Svoboda, D., Wolterink, J.M., Escobar, M. (eds.) SASHIMI 2022. LNCS, vol. 13570, pp. 79–90. Springer International Publishing, Cham (2022). https://doi.org/10.1007/978-3-031-16980-9_8
12. Foroozandeh, M., Eklund, A.: Synthesizing brain tumor images and annotations by combining progressive growing GAN and SPADE (2020). https://doi.org/10.48550/arxiv.2009.05946, https://arxiv.org/abs/2009.05946v1
13. Hoogeboom, E., Heek, J., Salimans, T.: simple diffusion: end-to-end diffusion for high resolution images
14. Goodfellow, I., Bengio, Y., Courville, A.: Deep Learning (2015). https://doi.org/10.1016/B978-0-12-391420-0.09987-X
15. Isensee, F., Jaeger, P.F., Kohl, S.A.A., Petersen, J., Maier-Hein, K.H.: nnU-Net: a self-configuring method for deep learning-based biomedical image segmentation. Nat. Methods **18**(2), 203–211 (2021). https://doi.org/10.1038/s41592-020-01008-z
16. Jack, C.R., et al.: The Alzheimer's Disease Neuroimaging Initiative (ADNI): MRI methods. J. Magn. Resonan. Imaging (JMRI) **27**(4), 685–691 (2008). https://doi.org/10.1002/JMRI.21049, https://pubmed.ncbi.nlm.nih.gov/18302232/
17. Jones, S., et al.: Cohort profile update: southall and brent revisited (SABRE) study: a UK population-based comparison of cardiovascular disease and diabetes in people of European, South Asian and African Caribbean heritage. Int. J. Epidemiol. **49**(5), 1441–1442 (2020). https://doi.org/10.1093/ije/dyaa135
18. Liu, L., Ren, Y., Lin, Z., Zhao, Z.: Pseudo numerical methods for diffusion models on manifolds (2022). https://doi.org/10.48550/arxiv.2202.09778, https://arxiv.org/abs/2202.09778v2
19. Menze, B.H., et al.: The multimodal brain tumor image segmentation benchmark (BRATS). IEEE Trans. Med. Imaging **34**(10), 1993–2024 (2015). https://doi.org/10.1109/TMI.2014.2377694, https://pubmed.ncbi.nlm.nih.gov/25494501, https://www.ncbi.nlm.nih.gov/pmc/articles/PMC4833122/
20. Park, T., et al.: Semantic image synthesis with spatially-adaptive normalization. In: Proceedings of IEEE CVPR, June 2019, pp. 2332–2341 (2019)

21. Pinaya, W.H.L., et al.: Brain Imaging Generation with Latent Diffusion Models. In: Mukhopadhyay, A., Oksuz, I., Engelhardt, S., Zhu, D., Yuan, Y. (eds.) DGM4MICCAI 2022. LNCS, vol. 13609, pp. 117–126. Springer Nature Switzerland, Cham (2022). https://doi.org/10.1007/978-3-031-18576-2_12

22. Qasim, A.B., et al.: Red-GAN: attacking class imbalance via conditioned generation. Yet another medical imaging perspective (2020). https://proceedings.mlr.press/v121/qasim20a.html

23. Rieke, N., et al.: The future of digital health with federated learning. NPJ Digit. Med. **3**(1), 119 (2020)

24. Salimans, T., Ho, J.: Progressive distillation for fast sampling of diffusion models

25. Sudre, C.H., Cardoso, M.J., Bouvy, W.H., Biessels, G.J., Barnes, J., Ourselin, S.: Bayesian model selection for pathological neuroimaging data applied to white matter lesion segmentation. IEEE Trans. Med. Imaging **34**(10), 2079–2102 (2015). https://doi.org/10.1109/TMI.2015.2419072

26. Tudosiu, P.D., et al.: Morphology-preserving autoregressive 3d generative modelling of the brain. In: Zhao, C., Svoboda, D., Wolterink, J.M., Escobar, M. (eds.) SASHIMI 2022. LNCS, vol. 13570, pp. 66–78. Springer, Cham (2022). https://doi.org/10.1007/978-3-031-16980-9_7

27. Wachinger, C., et al.: BrainPrint: a discriminative characterization of brain morphology. NeuroImage **109**, 232–248 (2015)

Rethinking a Unified Generative Adversarial Model for MRI Modality Completion

Yixiao Yuan[1], Yawen Huang[2], and Yi Zhou[1(✉)]

[1] School of Computer Science and Engineering, Southeast University, Nanjing, China
yxyuan@seu.edu.cn
[2] Tencent Jarvis Lab, Beijing, China

Abstract. Multi-modal MRIs are essential in medical diagnosis; however, the problem of missing modalities often occurs in clinical practice. Although recent works have attempted to extract modality-invariant representations from available modalities to perform image completion and enhance segmentation, they neglect the most essential attributes across different modalities. In this paper, we propose a unified generative adversarial network (GAN) with pairwise modality-shared feature disentanglement. We develop a multi-pooling feature fusion module to combine features from all available modalities, and then provide a distance loss together with a margin loss to regularize the symmetry of features. Our model outperforms the existing state-of-the-art methods for the missing modality completion task in terms of the generation quality in most cases. We show that the generated images can improve brain tumor segmentation when the important modalities are missing, especially in the regions which need details from various modalities for accurate diagnosis.

Keywords: Modality completion · Unified GAN · Missing-domain segmentation

1 Introduction

Magnetic resonance imaging (MRI) provides a range of imaging contrasts (e.g., T1, T1ce, T2 and FLAIR) for diagnosis of brain tumor, in which the tumor boundaries can be identified by comparing different modalities, as they provide complementary features [20]. To improve the performance of brain tumor segmentation, previous works [11,18] use all the modalities simultaneously. However, such a treatment is inappropriate due to practical limitations, e.g., long scanning time and image corruption [22]. To solve this problem, a direct way is to train a series of independent models dedicated to each missing situation [8,9,12,19,20,23]. Van Tulder et al. [19], Liu et al. [12] and Hu et al. [9] synthesized the missing modalities with available ones using a task-specific model. Wang et al. [20], Hu et al. [8] and Yang et al. [23] used knowledge distillation to

© The Author(s), under exclusive license to Springer Nature Switzerland AG 2024
A. Mukhopadhyay et al. (Eds.): DGM4MICCAI 2023, LNCS 14533, pp. 143–153, 2024.
https://doi.org/10.1007/978-3-031-53767-7_14

transfer knowledge from the full modality model to a modality-specific model. However, all these methods need more than one model, which are complicated and time-consuming.

Another solution is to build a unified framework for all possible missing situations during inference [3–7,24]. Havaei et al. [7] and Dorent et al. [6] fused multi-modal information by computing mean and variance across individual features. Instead of fusing the layers by computing mean and variance, Chen et al. [4] fused each modality into a modality-invariant representation model which gains robustness to missing modality data by feature disentanglement. Zhou et al. [24] introduced a correlation approach to exploring the latent multi-source correlation representation. In addition to making the model robust against missing situation, designing a unified image synthesis model [3,14,16,17] achieves image completion purpose, in which the complementary imaging modalities are available during inference. Shen et al. [17] and Ouyang et al. [14] used representation disentanglement to learn the modality-invariant representations and the modality-related representations and further use the modality-invariant representations to perform image synthesis.

The main motivation for our work lies in the limitations of acquiring a full battery of multi-modal MRI data, for example, high examination cost, restrictive availability of scanning time and image corruptions frequently crop up, especially in large scale studies. Although the existing works attempt to solve this issue by learning modality-invariant features shared by all available modalities for synthesizing missing data, the whole learning procedure overlooks the pairwise modality-shared features (which are exclusively shared between each pair of two modalities) resulting in inefficient usage of information as omission of real matched information. It is precisely this information can be used to identify tumor boundaries and improve the accuracy of segmentation. Our method addresses this limitation by explicitly learning the pairwise modality-shared features for each modality. By constructing the pairwise modality-shared feature disentanglement, our model can better capture the complementary information from multiple modalities and generate more accurate and diverse data for missing modalities.

Specifically, we propose a novel pairwise modality-shared feature disentanglement method by building a unified synthesis model to generate missing modalities. **The main contributions of this paper are summarized as follows:**

- We propose a novel unified GAN-based framework for random missing modality completion with the representation disentanglement, i.e., learning the T1-shared, T1ce-shared, T2-shared and FLAIR-shared features for each modality.
- Experimental results show that the proposed method achieves better performance in most cases than the previous approaches of multi-domain MRI image completion.
- We demonstrate that the generated multi-modality data can improve the performance of brain tumor segmentation.

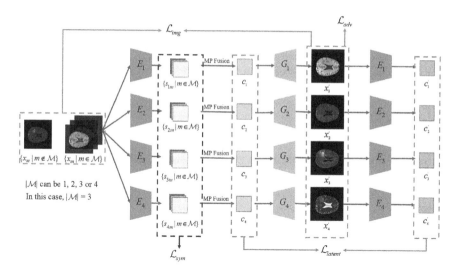

Fig. 1. The proposed unified GAN. The number of available modalities $|\mathcal{M}| = 3$ is given in this example. Available modalities are disentangled into pairwise modality-shared features one by one, where c_i is computed by multi-pooling fusion (MP Fusion) of all the extracted pairwise modality-shared features $\{s_{im} \mid m \in \mathcal{M}\}$ extracted by the encoder E_i. These networks are then trained by the reconstruction loss $\mathcal{L}_{img}$, latent consistency loss $\mathcal{L}_{latent}$, symmetrical loss $\mathcal{L}_{sym}$ and adversarial loss $\mathcal{L}_{adv}$.

2 Methods

2.1 Pairwise Modality-Shared Feature Disentanglement

To complete missing modalities from available modalities, as shown in Fig. 1, our model contains pairwise modality-specific encoders E_i to extract the features $s_{ij} = E_{ij}(x_j)$ shared in modality i from any modality input x_j ($1 \leq i \leq M, 1 \leq j \leq M$), where M is the number of modalities with $M = 4$ in this task. For each modality, we extract M pairwise modality-shared features including itself and we perform image translation according to the pairwise modality-shared features. Specifically, we want to learn the T1-shared, T1ce-shared, T2-shared and FLAIR-shared features for each available modality and fuse features from available modalities to do modality completion. The generator G_i reconstructs x_i from the fused pairwise modality-shared features $c_i = Fusing\{s_{ij} \mid j \in \mathcal{M}\}$ extracted from available modality set $\mathcal{M}$.

Conditional Modality Encoder. Considering that each modality data is encoded by an independent encoder, generating M output modalities from M input modalities requires M^2 encoders with pair-wise learning, which is too complex and difficult to implement. Inspired by [14], we use M conditional modality encoders composed by conditional convolution (CondConv) [21]. The parameters of CondConv are decided by the input modality i using a mixture-of-experts model $CondConv(x; i) = \sigma\left(\left(\beta_1^i \cdot W_1 + \ldots + \beta_n^i \cdot W_n\right) \circledast x\right)$, where $\sigma(\cdot)$ is the sigmoid activation function, $\circledast$ denotes a regular convolution, $\{W_1, \ldots, W_n\}$ are the learnable kernels associated with n experts, and $\{\beta_1^i, \ldots \beta_n^i\}$ are the modality-specific mixture weights. Therefore, for each modality, e.g., modality j, the set of independent encoders $\{E_{ij}(x_i)\}$ is coupled into one encoder $E_j(x_i; i)$, and the training data to each encoder increased M-fold, which makes the model more robust and easier to train.

Multi-pooling Feature Fusion. In modality-missing situations, suppose available modalities set are $\mathcal{M}$, and for the target modality we want to translate into, we will get $|\mathcal{M}|$ pairwise modality-shared features $\{s_1, \ldots, s_{|\mathcal{M}|}\}$, which is not fixed.

To fuse all features into a fixed size as the input for decoders, we first concatenate the output of pooling functions $MaxPool$, $MeanPool$ and $MinPool$ to keep as much information as possible. Then, we use 1×1 convolution to make the number of channels equal to the output of encoders. Compared with a single pooling function, this multi-pooling fusion method can keep different perspective features, which allows our fused features to provide more details to the following network. For the fusion operation, although the number of available modalities is randomly, the input size of decoders is the same as the output of the encoders, and thus we can perform a latent consistency loss.

2.2   Training Objectives

Image Consistency Loss. Generator G_i is supposed to synthesize an image that is similar to the input image $x_i \sim \mathcal{X}_i$. To make the generated images similar to our targets, we employ an image consistency loss as:

$$\mathcal{L}_{img} = \sum_{i=1}^{M} \mathbb{E}_{x_i \sim \mathcal{X}_i} \left[\|G_i(c_i) - x_i\|_1\right], \tag{1}$$

where $c_i = Fusing\{E_i(x_j; j) \mid j \in \mathcal{M}\}$ fuses modality-i-shared features from the available modality set $\mathcal{M}$. Here, we use the L_1 loss to strengthen structure related generation.

Latent Consistency Loss. The latent consistency loss is another common loss used in image-to-image translation, which encourages the features derived from raw inputs to be similar to the ones from the synthesized images,

$$\mathcal{L}_{latent} = \sum_{i=1}^{M} \mathbb{E}_{x_i \sim \mathcal{X}_i} \left[\|E_i(G_i(c_i); i) - c_i\|_1\right]. \tag{2}$$

Specifically, the fused modality-i-shared features c_i are the target to be recovered by our encoder E_i.

Adversarial Loss. To minimize the difference between the distributions of generated images and real images, we define the adversarial loss as:

$$\mathcal{L}_{adv} = \sum_{i=1}^{M} \mathbb{E}_{x_i \sim \mathcal{X}_i} \left[\log \left(1 - D_i \left(G_i \left(c_i \right) \right) \right) + \log D_i \left(x_i \right) \right], \tag{3}$$

where D_i is the discriminator for modality i to distinguish the generated images.

Symmetrical Loss. An ideal pairwise modality-shared feature is symmetrical, for example, T1-shared features extracted from T2 should be similar to T2-shared features extracted from T1. To enforce that pairwise modality-shared features are disentangled well, we add the similarity regularization as:

$$\mathcal{L}_{sym} = \sum_{i \in \mathcal{M}} \sum_{j \in \mathcal{M}} \sum_{k=1, k \neq i,j}^{M} \mathbb{E} \left[\| s_{ij} - s_{ji} \|_2 + \max(0, \alpha - d(s_{ij}, s_{ji}) + d(s_{ij}, s_{ik}) \right], \tag{4}$$

where $d(\cdot, \cdot)$ calculates the L_2 distance between two tensors and $s_{ij} = E_i(x_j; j)$ denotes modality-i-shared features extracted from modality j. The L_2 loss term encourages pairwise modality-shared features between the same two modalities to be similar, and the margin loss prevents the encoder from mapping all pairwise modality-shared features into the same location in the feature space.

Total Loss. The encoders, generators and discriminators are jointly trained to optimize the total objective function as:

$$\mathcal{L} = \lambda_{img} \mathcal{L}_{img} + \lambda_{latent} \mathcal{L}_{latent} + \lambda_{adv} \mathcal{L}_{adv} + \lambda_{sym} \mathcal{L}_{sym}. \tag{5}$$

The ultimate goal of the overall co-training procedure is to optimize the function via $\min_{E,G} \max_D \mathcal{L}$.

3   Experiments

3.1   Experimental Settings

Dataset and Preprocessing. We evaluate our proposed method with the multimodal brain tumor segmentation dataset (BraTS 2018) [1,2,13], which provides multi-modal brain MRI with four modalities: T1, T1ce, T2, and FLAIR. We divide regions into three parts: whole tumor (WT), tumor core (TC), and enhancing tumor (ET) according to the provided labels. Specifically, 200, 27 and 58 subjects are randomly selected for training, validation and testing and then we select the middle axial slices of each MRI volume and discarding the volumes with fewer than a certain threshold of valid pixels. We normalize the intensity of each slice to $[-1, 1]$. A patch of size 224×224 is randomly cropped during training as input to the network.

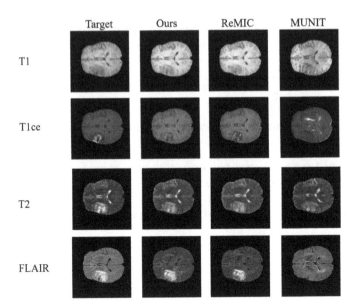

Fig. 2. Comparison of generated images on BraTS with a single missing modality. The missing modalities are shown in rows, while the compared methods are shown in columns.

Implementation Details. Experiments are implemented in Pytorch and performed on a single NVIDIA RTX3090 GPU. In all experiments, we set $\lambda_{img} = 10$, $\lambda_{latent} = 1$, $\lambda_{adv} = 1$, $\lambda_{sym} = 1$ and $\alpha = 0.1$ in $\mathcal{L}_{sym}$.

The proposed encoder includes a down-sampling procedure along with residual blocks, which are composed of convolutional blocks. Specifically, a 7×7 conditional convolutional block with stride equal to 1 and two 4×4 conditional convolutional blocks with stride equal to 2 are involved in our down-sampling module. In total, four residual blocks are included, in which each block has two convolutional blocks with size of 3×3, 256 filters and stride equal to 1. The decoder contains four residual blocks, where each one has two convolutional blocks with size of 3×3, 256 filters and stride equal to 1. To up-sample the fused features matching to the original image size, two nearest-neighbor upsampling layers together with a 5×5 convolutional block having stride equal to 1 are employed, where the filter numbers are from 64 to 128 to 256 to 128 to 64. The reconstructed image can then be generated using a 7×7 convolutional block with stride equal to 1 combining with a filter. As for the discriminator, it has four convolutional blocks with size of 4×4, and sets stride of 2, with filter numbers ranging from 64 to 128 to 256 to 512. We use leaky ReLU with slope of 0.2 in the discriminator. During training, the number of available modalities is random, except 0, and for each missing situation, the available modalities are also randomly distributed.

Table 1. Comparison with the n-to-n image translation method ReMIC [17] for the different combinations of available modalities denoted by ✓.

FLAIR	T1	T1ce	T2	ReMIC (PSNR)	Ours (PSNR)	ReMIC (SSIM)	Ours (SSIM)
✓	✓	✓		27.00	**27.68**	0.915	**0.923**
✓	✓		✓	26.83	**27.29**	0.895	**0.909**
✓		✓	✓	26.68	**27.01**	0.918	**0.925**
	✓	✓	✓	**25.30**	25.23	**0.874**	0.870
		✓	✓	**25.67**	25.61	**0.892**	0.891
	✓		✓	25.94	**25.99**	0.884	**0.887**
	✓	✓		25.25	**25.50**	0.878	**0.880**
✓			✓	25.62	**25.94**	0.893	**0.899**
✓		✓		26.50	**26.94**	0.901	**0.913**
✓	✓			26.54	**27.05**	0.899	**0.910**
✓				24.98	**25.18**	0.872	**0.875**
	✓			25.12	**25.31**	0.875	**0.878**
		✓		24.89	**25.13**	**0.875**	0.875
			✓	24.57	**24.60**	0.870	**0.872**
Mean				25.78	**26.03**	0.889	**0.893**

Table 2. Comparison with the 1-to-1 image translation method MUNIT [10]. For 1-to-1 case, we use the same modality as MUNIT to perform image translation. For n-to-1 case, our method uses all the other modalities to generate the missing modality data.

Missing modality	MUNIT PSNR	MUNIT SSIM	Ours (1-to-1) PSNR	Ours (1-to-1) SSIM	Ours (n-to-1) PSNR	Ours (n-to-1) SSIM
T1	21.23	0.865	25.75	0.902	**27.01**	**0.925**
T1ce	22.99	0.867	26.25	0.892	**27.29**	**0.909**
T2	22.47	0.855	25.51	0.893	**27.68**	**0.923**
FLAIR	20.22	0.807	23.49	0.844	**25.23**	**0.870**

Baseline Methods. [1] To evaluate the performance of our model, we compare it with the existing state-of-the-art image translation methods, i.e., MUNIT [10] and ReMIC [17]. MUNIT is a 1-to-1 image translation method between two domains through representation disentanglement. We use T1 data to generate other modalities data, and FLAIR data to generate T1 images. ReMIC builds a unified n-to-n image translation like our method. However, **ReMIC uses modality-invariant feature disentanglement while our method is based on pairwise modality-shared feature disentanglement.**

---

[1] Due to the scarcity of works in unified modality completion and lack of open-source implementations, we were only able to compare our method against few approaches.

3.2   Results and Analysis

Evaluation on Image Completion. We evaluate the methods on the missing modality completion task with different datasets. Figure 2 shows the qualitative results on a single missing modality. Both ReMIC and ours generate the missing modalities with a better quality, considering the incorporated complementary information from multiple available modalities. Comparing with ReMIC, our method keeps more details, e.g., the tumor region boundary is clearer, which is achieved by learning pairwise modality-shared features in each available modality. Besides, artificial details often presented in the MUINT results do not appear in ours. Furthermore, as shown in Table 1 abd Table 2, our method outperforms the baselines in terms of peak-signal-noise ratio (PSNR) and structural similarity index (SSIM) in most cases, which suggests that our method produces more realistic MRI images.

Table 3. The performance of segmentation under missing-modality situation (Dice %).

Missing modality	Tumor Core		Whole Tumor		Enhancing	
	ReMIC	Ours	ReMIC	Ours	ReMIC	Ours
None	75.5		89.8		73.2	
T1	73.8	**74.6**	**89.3**	89.2	71.9	**72.5**
T1ce	49.8	**54.6**	87.5	**87.8**	14.8	**25.2**
T2	75.1	**75.2**	**88.6**	87.8	71.8	**72.0**
FLAIR	74.2	**75.7**	85.3	**87.2**	71.3	**72.9**

Table 4. Effectiveness of each module on missing modality completion (PSNR in dB).

Methods	Missing modality			
	T1	T1ce	T2	FLAIR
Baseline	26.32	26.57	26.89	24.84
Baseline + CME	26.88	26.93	27.30	25.11
Baseline + CME + MP Fusion	26.94	27.04	27.38	25.18
Baseline + CME + MP Fusion + $\mathcal{L}_{sym}$	**27.01**	**27.29**	**27.68**	**25.23**

Evaluation on Segmentation. Segmentation with missing modalities can be solved by our method. We evaluate the segmentation performance in missing situations with Dice coefficient on U-Net [15], pre-trained on a complete dataset with all modality images. The result is shown in Table 3. None means performing segmentation without any missing modalities. We observe that our method performs well in segmenting tumor core and enhancing tumor, while ReMIC is only competitive in segmenting whole tumor. The whole tumor region can be distinguished in all the modalities, and in this way it belongs to the modality-invariant features used in ReMIC. Through pairwise modality-shared feature disentanglement, our method can provide more information than mere modality-invariant features. Particularly for enhancing tumor and distinguishing tumor core with more modality-specific information, our method shows better performance.

Ablation Study. In addition, we investigate the effectiveness of each component in our method in single modality completion. We first set up a baseline network with the traditional convolution and use $MaxPool$ for fusion. Then, we add the conditional modality encoder (CME), the multi-pooling feature fusion (MP Fusion), and the symmetrical loss one by one into the baseline network. The results are shown in Table 4. From Table 4, we note that the qualities of generations for different missing modality data are improved simultaneously, which proves the superiority of the proposed fusion module and the symmetrical loss.

4   Conclusions

In this work, we proposed a unified GAN-based network by investigating our pairwise modality-shared features between modalities, rather than directly using modality-invariant features for missing modality completion. We introduced a conditional modality encoder, a multi-pooling feature fusion method, and a symmetrical loss to improve the model performance. Experimental results illustrated that the proposed method achieves better performance than the previous approaches in multi-domain MRI image completion. In addition, we demonstrated that our generated images can improve the performance of the downstream task like brain tumor segmentation, in missing situations, especially in the region which needs details from different modalities for accurate analysis.

References

1. Bakas, S., et al.: Advancing the cancer genome atlas glioma MRI collections with expert segmentation labels and radiomic features. Sci. Data **4**(1), 1–13 (2017)
2. Bakas, S., et al.: Identifying the best machine learning algorithms for brain tumor segmentation, progression assessment, and overall survival prediction in the BRATS challenge. arXiv preprint arXiv:1811.02629 (2018)
3. Chartsias, A., Joyce, T., Giuffrida, M.V., Tsaftaris, S.A.: Multimodal MR synthesis via modality-invariant latent representation. IEEE Trans. Med. Imaging **37**(3), 803–814 (2017)
4. Chen, C., Dou, Q., Jin, Y., Chen, H., Qin, J., Heng, P.-A.: Robust multimodal brain tumor segmentation via feature disentanglement and gated fusion. In: Shen, D., Liu, T., Peters, T.M., Staib, L.H., Essert, C., Zhou, S., Yap, P.-T., Khan, A. (eds.) MICCAI 2019. LNCS, vol. 11766, pp. 447–456. Springer, Cham (2019). https://doi.org/10.1007/978-3-030-32248-9_50

5. Ding, Y., Yu, X., Yang, Y.: RFNet: region-aware fusion network for incomplete multi-modal brain tumor segmentation. In: Proceedings of the IEEE/CVF International Conference on Computer Vision, pp. 3975–3984 (2021)
6. Dorent, R., Joutard, S., Modat, M., Ourselin, S., Vercauteren, T.: Hetero-modal variational encoder-decoder for joint modality completion and segmentation. In: Shen, D., et al. (eds.) Hetero-modal variational encoder-decoder for joint modality completion and segmentation. LNCS, vol. 11765, pp. 74–82. Springer, Cham (2019). https://doi.org/10.1007/978-3-030-32245-8_9
7. Havaei, M., Guizard, N., Chapados, N., Bengio, Y.: HeMIS: hetero-modal image segmentation. In: Ourselin, S., Joskowicz, L., Sabuncu, M.R., Unal, G., Wells, W. (eds.) MICCAI 2016. LNCS, vol. 9901, pp. 469–477. Springer, Cham (2016). https://doi.org/10.1007/978-3-319-46723-8_54
8. Hu, M., et al.: Knowledge distillation from multi-modal to mono-modal segmentation networks. In: Martel, A.L., et al. (eds.) MICCAI 2020. LNCS, vol. 12261, pp. 772–781. Springer, Cham (2020). https://doi.org/10.1007/978-3-030-59710-8_75
9. Hu, X., Shen, R., Luo, D., Tai, Y., Wang, C., Menze, B.H.: AutoGAN-synthesizer: neural architecture search for cross-modality MRI synthesis. In: International Conference on Medical Image Computing and Computer Assisted Intervention. pp. 397–409. Springer Nature Switzerland, Cham (2022)
10. Huang, X., Liu, M.Y., Belongie, S., Kautz, J.: Multimodal unsupervised image-to-image translation. In: Proceedings of the European Conference on Computer Vision, pp. 172–189 (2018)
11. Isensee, F., Jäger, P.F., Full, P.M., Vollmuth, P., Maier-Hein, K.H.: nnU-net for brain tumor segmentation. In: Crimi, A., Bakas, S. (eds.) BrainLes 2020. LNCS, vol. 12659, pp. 118–132. Springer, Cham (2021). https://doi.org/10.1007/978-3-030-72087-2_11
12. Liu, Z., Wei, J., Li, R.: TFusion: Transformer based N-to-One Multimodal Fusion Block. arXiv preprint arXiv:2208.12776 (2022)
13. Menze, B.H., et al.: The multimodal brain tumor image segmentation benchmark (BRATS). IEEE Trans. Med. Imaging 34(10), 1993–2024 (2014)
14. Ouyang, J., Adeli, E., Pohl, K.M., Zhao, Q., Zaharchuk, G.: Representation disentanglement for multi-modal brain MRI analysis. In: Feragen, A., Sommer, S., Schnabel, J., Nielsen, M. (eds.) IPMI 2021. LNCS, vol. 12729, pp. 321–333. Springer, Cham (2021). https://doi.org/10.1007/978-3-030-78191-0_25
15. Ronneberger, O., Fischer, P., Brox, T.: U-Net: convolutional networks for biomedical image segmentation. In: Navab, N., Hornegger, J., Wells, W.M., Frangi, A.F. (eds.) MICCAI 2015. LNCS, vol. 9351, pp. 234–241. Springer, Cham (2015). https://doi.org/10.1007/978-3-319-24574-4_28
16. Sharma, A., Hamarneh, G.: Missing MRI pulse sequence synthesis using multi-modal generative adversarial network. IEEE Trans. Med. Imaging 39(4), 1170–1183 (2019)
17. Shen, L., et al.: Multi-domain image completion for random missing input data. IEEE Trans. Med. Imaging 40(4), 1113–1122 (2020)
18. Tseng, K.L., Lin, Y.L., Hsu, W., Huang, C.Y.: Joint sequence learning and cross-modality convolution for 3D biomedical segmentation. In: Proceedings of the IEEE Conference on Computer Vision and Pattern Recognition (CVPR), pp. 6393–6400 (2017)
19. van Tulder, G., de Bruijne, M.: Why does synthesized data improve multi-sequence classification? In: Navab, N., Hornegger, J., Wells, W.M., Frangi, A.F. (eds.) MICCAI 2015. LNCS, vol. 9349, pp. 531–538. Springer, Cham (2015). https://doi.org/10.1007/978-3-319-24553-9_65

20. Wang, Y., et al.: ACN: adversarial co-training network for brain tumor segmentation with missing modalities. In: de Bruijne, M., et al. (eds.) MICCAI 2021. LNCS, vol. 12907, pp. 410–420. Springer, Cham (2021). https://doi.org/10.1007/978-3-030-87234-2_39

21. Yang, B., Bender, G., Le, Q.V., Ngiam, J.: Condconv: conditionally parameterized convolutions for efficient inference. Adv. Neural. Inf. Process. Syst. **32**, 1307–1318 (2019)

22. Yang, H., Sun, J., Yang, L., Xu, Z.: A unified hyper-GAN model for unpaired multi-contrast MR image translation. In: de Bruijne, M., Cattin, P.C., Cotin, S., Padoy, N., Speidel, S., Zheng, Y., Essert, C. (eds.) MICCAI 2021. LNCS, vol. 12903, pp. 127–137. Springer, Cham (2021). https://doi.org/10.1007/978-3-030-87199-4_12

23. Yang, Q., Guo, X., Chen, Z., Woo, P.Y.M., Yuan, Y.: D2-Net: dual disentanglement network for brain tumor segmentation with missing modalities. IEEE Trans. Med. Imaging **41**(10), 2953–2964 (2022)

24. Zhou, T., Canu, S., Vera, P., Ruan, S.: Brain tumor segmentation with missing modalities via latent multi-source correlation representation. In: Martel, A.L., et al. (eds.) MICCAI 2020. LNCS, vol. 12264, pp. 533–541. Springer, Cham (2020). https://doi.org/10.1007/978-3-030-59719-1_52

Diffusion Models for Generative Histopathology

Niranjan Sridhar[1]([✉]) [ID], Michael Elad[2] [ID], Carson McNeil[1] [ID], Ehud Rivlin[2], and Daniel Freedman[2] [ID]

[1] Verily Research, South San Francisco 94080, USA
nirsd@verily.com
[2] Verily Research, Tel Aviv, Israel

Abstract. Conventional histopathology requires chemical staining to make tissue samples usable by pathologists for diagnosis. This introduces cost and variability and does not conserve the tissue for advanced molecular analysis of the sample. We demonstrate the use of conditional denoising diffusion models applied to non-destructive autofluorescence images of tissue samples in order to generate virtually stained images. To demonstrate the power of this technique, we would like to measure the perceptual quality of the generated images; however, standard measures like the Frechet Inception Distance (FID) are inappropriate for this task, as they have been trained on natural images. We therefore introduce a new perceptual measure, the Frechet StainNet Distance (FSD), and show that our model attains significantly higher FSD than competing pix2pix models. Finally, we also present a method of quantifying uncertain regions of the image using the variations produced by diffusion models.

Keywords: Diffusion · Pathology

1 Introduction

Conventional histopathology involves obtaining tissue sections from patient biopsies and applying chemical staining protocols which highlight different biological features of the tissue. This stained tissue can then be assessed and diagnosed by pathologist using a brightfield (BF) microscope. There are many chemical stains corresponding to different features to be highlighted. However, the process of staining can be destructive. A given stained tissue sample often cannot be used again for other analyses. Therefore, the cost of advanced testing, research or second opinions, which are often required for newer/rarer diseases, can be prohibitive. Additional drawbacks of histochemical staining include expensive laboratory infrastructure, slow processing times and the inherent variability in equipment and expertise.

Virtual staining [3, 15, 18, 19] is an AI-enabled alternative which removes the need for chemical staining. Tissue samples are imaged using a non-destructive auto-fluorescence (AF) scanner. The AF image records the spatial distribution

© The Author(s), under exclusive license to Springer Nature Switzerland AG 2024
A. Mukhopadhyay et al. (Eds.): DGM4MICCAI 2023, LNCS 14533, pp. 154–163, 2024.
https://doi.org/10.1007/978-3-031-53767-7_15

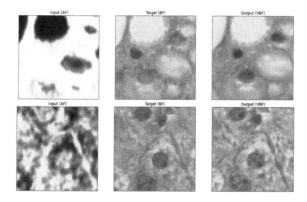

Fig. 1. Diffusion models are able to generate virtual stained outputs with both high fidelity to the target image and high perceptual quality.

emission spectra of the tissue after exposing it to excitation lasers and therefore contains information about both the condition and location of different biological features of the tissue. An image-to-image translation model can then be used to learn the mapping from the AF image of the tissue to its stained BF image. If this virtually stained image can capture all of the clinical features of real stained tissues, the pathologist can use the translated image for clinical diagnosis. Since the same AF image can be used for any number of stain types and the original tissue is preserved, virtual staining can greatly reduce the cost and effort of clinical pathology. The crucial step in this process is the image-to-image translation algorithm. In this paper, we apply conditional diffusion models to this task. We make the following key contributions:

1. **Diffusion Models for Staining.** We present a conditional diffusion model for virtual staining, which maps AF images to chemically stained BF images.
2. **Frechet StainNet Distance (FSD).** We develop a new technique for evaluating the perceptual quality of our output, referred to as the Frechet StainNet Distance. As compared to FID, FSD is much more appropriate for evaluating stained microscopy images. We note that FSD may be applied to other scenarios beyond that described in this paper.
3. **Significantly Improved Perceptual Quality.** We show empirically that as compared to conditional GANs, the diffusion models perform significantly better on perceptual quality as measured by FSD, while remaining comparable on distortion measures.
4. **Uncertainty Quantification.** We use the capabilities of our diffusion method to provide a reliable approximation of the uncertainty associated with the stained estimate per each pixel (Fig. 1).

2   Related Work

Virtual Staining: Until recently, the state-of-the-art in image-to-image translation were conditional GANs such as pix2pix [11] for paired datasets, and

CycleGANs [26] for unpaired ones. To prevent the GAN generator from hallucinating realistic images unrelated to the input, pixel-wise losses such as L_p distance between the virtual images and their corresponding ground truth are also used in addition to adversarial losses. A major drawback of GANs is that they are hard to train due to loss instability and mode collapse, e.g. see [2]. Previous efforts in virtual staining have used both pix2pix and CycleGANs [3,15,18,19]. These prior work often only report distortion measures such as L_p norms. However, for virtual stains to replace chemical stains in a clinical workflow, they must also look similar to human pathologists. Therefore it is important to benchmark these models on perceptual quality. In this work, we benchmark our models against a pix2pix model inspired by Rivenson et al. [19] with 128×128 resolution inputs, two discriminator losses (conditional and unconditional) and two pixel-wise losses (L1 and rotated L1).

Diffusion: Diffusion models [22] have recently emerged with impressive results on the task of unconditional image generation, beating GANs for generating images with high diversity and perceptual quality [6,8]. An important variation of these techniques is the conditional diffusion model, see e.g. [9,21,23,25] which is the basis of our current work. Saharia et al. [9,21] show that diffusion models can produce images with high perceptual quality without losing the structural and semantic information of the input image on a number of image-to-image translation tasks.

Perceptual Measures: The FID score [7] is commonly used to quantify perceptual quality. However, since the standard InceptionV3 model [24] used in FID has been trained on natural images, the measure is likely to have difficulty differentiating between varying distributions of histological images, which can be close to each other in the space of natural images. This has been documented previously for other data types such as audio and molecular data [12,16]. The tradeoff between the distortion between the expected and the predicted images, and the perceptual quality of the predicted image has been well studied [5]. Regression models that minimize the distortion between the labels and the prediction cannot produce outputs belonging to the expected output distribution. Therefore such models have low perceptual quality, i.e. they do not produce images that look like real images to humans. In contrast, GANs and diffusion models have the ability to generate images of high perceptual quality.

Uncertainty Quantification: Quantifying uncertainty in deep learning is difficult due to the lack of a closed form expression for the density. Deterministic models that produce a single output per input require complex interrogation to extract such information. Perturbative methods, such as LIME [17], involve repeated inference with varying data augmentations to estimate the effect of input variations on the output. In contrast, integrated methods, such as quantile regression [13], involve adding credible interval bound estimation as an additional training objective, either during the original training or after it. Generative models provide a new opportunity as they can sample different outputs from the target distribution upon repeated inference. Following [10], in conditional diffusion models we apply a series of inference rounds and generate multiple results to approximate the distribution of the output conditioned on the input.

3   Methodology

3.1   Denoising Diffusion Probabilistic Models

A denoising diffusion probabilistic model [8] can be described as a parameterized Markov chain. The forward diffusion process is a series of steps that add small amounts of Gaussian noise to the data until the signal is destroyed. Given data x_0 which we consider a sampling of the distribution $q(x_0)$, we can create T vectors $\{x_1, ..., x_T\}$ of the same dimensions as x_0 defined by the forward diffusion process:

$$q(x_t|x_{t-1}) = \mathcal{N}(\sqrt{1 - \beta_t}x_{t-1}, \beta_t \mathbf{I}) \text{ for } t = 1, 2, \ ... \ , T,$$

i.e. x_t is constructed as a mixture of x_{t-1} with a Gaussian noise, with the scaling variance parameter $\beta_t \in (0, 1)$. It immediately follows from the above:

$$q(x_t|x_0) = \prod_{i=1}^{t} q(x_i|x_{i-1}) = \mathcal{N}(\sqrt{\alpha_t}x_0, (1 - \alpha_t)\mathbf{I}) \text{ for } t = 1, 2, \ ... \ , T,$$

where $\alpha_t = \prod_{i=1}^{t}(1 - \beta_i)$. The number of steps T and the variance schedule β_t are chosen such that x_T is pure Gaussian noise, while at the same time the variances β_t of the forward process are small. Under these conditions, we can learn a reverse process p_θ which can be defined as

$$p_\theta(x_{t-1}|x_t) = \mathcal{N}(\mu_\theta(x_t, t), \sigma_\theta(x_t, t)) \text{ for } t = T, T - 1, \ ... \ , 1. \tag{1}$$

Note that chaining these probabilities leads to a sampled outcome x_0 following the probability density function

$$p_\theta(x_0) = p(x_T) \prod_{t=T}^{1} p_\theta(x_{t-1}|x_t).$$

Returning to our goal of reversing the diffusion process, we can leverage the following relationship:

$$q(x_{t-1}|x_t, x_0) = \mathcal{N}\left(\frac{\sqrt{\alpha_{t-1}}\beta_t}{1 - \alpha_t}x_0 + \frac{\sqrt{1 - \beta_t}(1 - \alpha_{t-1})}{1 - \alpha_t}x_t, \frac{1 - \alpha_{t-1}}{1 - \alpha_t}\beta_t \mathbf{I}\right) \tag{2}$$

Observe the similarity between Eq. (1) and the above expression, where the later adds the knowledge of x_0. Thus, we can approximate p_θ by aligning the two moments of these Gaussians, which imply that we use a learned denoiser neural network $T_\theta(x_t, t)$ for estimating x_0 from x_t and t:

$$\mu_\theta(x_t, t) = \frac{\sqrt{\alpha_{t-1}}\beta_t}{1 - \alpha_t}T_\theta(x_t, t) + \frac{\sqrt{1 - \beta_t}(1 - \alpha_{t-1})}{1 - \alpha_t}x_t. \tag{3}$$

During inference, this denoising neural network is recursively applied, starting from pure Gaussian noise, to produce samples from the data distribution.

In conditional diffusion, the denoiser is modified to include AF images of tissue sample, y, concatenated to the input, both during training and inference. As a result, with the modified denoiser $T_\theta(x_t, y, t)$, the output of the diffusion model is a sample from the posterior data distribution $q(x_0|y)$. This allows conditional diffusion models to be used for image-to-image translation where the input image is used as the condition.

3.2   Architecture

As is common practice in the diffusion literature [21], rather than learning $T_\theta(x_t, t)$ which returns the clean signal, one learns the noise itself (which is trivially related to the clean signal). To learn this noise estimator, we adopt the UNet [20] denoiser architecture, as proposed by Ho et al. [8] for diffusion models, and the improvements proposed by Saharia et al. [21]. The UNet model uses a stack of 6 blocks, each made of 2 residual layers and 1 downsampling convolution, followed by a stack of 6 blocks of 2 residual layers and 1 upsampling convolution. Skip connections connect the layers with the same spatial size. In addition, we use a global attention layer with 2 heads at each downsampled resolution and add the time-step embedding into each residual block.

3.3   Perceptual Quality Measures

For each model, we run inference on 20,000 128×128 tiles and evaluate the virtual stain results against real stain patches. In addition to standard L_p-based distortion measures, we consider the following two measures of perceptual quality:

FID: The Frechet Inception Distance (FID) score is a perceptual measure shown to correlate well with human perception of realism [7]. FID measures the Frechet distance between two multivariate Gaussians fit to features of generated and real images extracted by the pre-trained InceptionV3 model [24].

FSD: We created a new custom measure to characterize the perceptual quality of stained images, which we dub the Frechet StainNet Distance (FSD). We create a dataset where each training example is a 128×128 patch of a stained BF images with a corresponding label representing the slide-level clinical Non-Alcoholic SteatoHepatitis (NASH) steatosis score (for more details on this score, see Sect. 4.1). We then train a classification model, StainNet, on this dataset. The features from StainNet are then taken to be the outputs of the penultimate layer of the StainNet network. Analogously to FID, FSD then measures the Frechet distance between two multivariate Gaussians fit to the StainNet features: the first Gaussian for the generated images and the second Gaussian for the real images. We note that FSD may be applied to other scenarios beyond that described in this paper.

Table 1. Quantitative evaluation results of different methods. All evals are done on 20,000 test image patches.

Model	FID	FSD	L1	L2
Regression	356.4	624.8	**10.5**	**13.8**
pix2pix	115.7	54.1	12.7	18.1
Diffusion-B	82.2	15.2	15.1	21.1
Diffusion-B/R	81.7	34.5	13.5	19.3
Diffusion-L	**69.1**	**4.5**	13.4	19.3

3.4 Sample Diversity and Uncertainty

To calculate pixel-wise 90% credible intervals (i.e. we expect 10% of samples to fall outside the bounds), we follow the approach proposed by Hoshen et al. [10]. We sample 20 outputs for every input image, and use these to approximate the output image distribution and its 5th and 95th quantiles as the bounds. The credible interval size is then the difference between the upper and the lower bound values for each pixel. This well-motivated but heuristic notion of uncertainty is then properly calibrated using a calibration factor λ to the interval bounds, which is determined using our validation set [1,4,10].

4 Experiments

4.1 Experimental Setup

Dataset. We use a proprietary dataset collected from a clinical study of patients diagnosed with Non-Alcoholic SteatoHepatitis (NASH). The dataset contains 192 co-registered pairs of images of whole slides of liver tissue: one AF image (26 spectral channels) and one H&E chemically stained BF image (3 RGB channels). The whole slides are captured at 40x resolution yielding large gigapixel images of variable shapes and sizes. We split the slides into train/val/test data in 0.5:0.2:0.3 ratio. Finally, we extract paired patches of size 128 × 128 from both AF and BF images, and all of the training and evaluation is done at the patch level. Each slide is between 1000 to 10000 pixels height and width and corresponds to between 700 and 3000 patches; thus the combined dataset is approximately 200,000 patches. In addition, for each slide we also have a clinical steatosis score. This score is an ordinal class between 0–3 assigned by human expert hepatopathologists quantifying the amount of liver disease features they observe in the whole slide.

Training. The diffusion model is trained on 16 TPUv3 cores in parallel. We use a batch size of 16 and a learning rate of 1e−5 throughout the entire training for 1.5 million steps or 120 epochs. We choose the number of diffusion steps $T = 1000$ and set the forward diffusion variances β_t to increase following a cosine function from $\beta_1 = 10^{-4}$ to $\beta_T = 0.02$, in accordance with the findings of Nichol and Dhariwal [14].

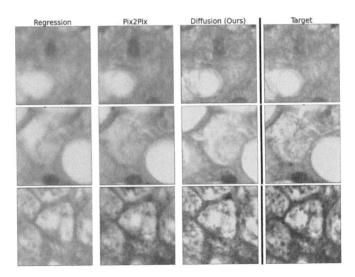

Fig. 2. We compare our Diffusion-L model with the target and benchmark it against regression and pix2pix models. Images generated by our diffusion model are closer to the target images in both texture and color.

Model Variants. In addition to our large diffusion model *Diffusion-L*, which has already been described, we train a number of variants. *Diffusion-B* is our base model which is similar to Diffusion-L but with only one single head attention layer at the 16×16 layer. The *Diffusion-B/R* model is the base model trained with an additional feature, in which a random part of the target image is masked and used as a prior; during inference, however, the prior is completely masked so that the generation is comparable to the other models.

4.2 Image Quality

We compare our diffusion model described above with a naive regression model, as well as a pix2pix (conditional GAN) model. Both models also use a UNet architecture, and the pix2pix model has additional unconditional and conditional adversarial losses. The results are presented in Table 1, which shows both the FID and FSD scores which measure perceptual quality, as well as L1 and L2 norms which measure distortion. Qualitative examples are shown in Fig. 2.

All of the diffusion models do score better (lower) in terms of FID scores; nevertheless, as previously noted, FID is not a very discriminative perceptual measure for stained pathology images, as it has been trained on natural images. For example, the Diffusion-B and Diffusion-B/R models attain almost identical FIDs. By contrast, FSD is much more discriminative and clearly shows that Diffusion-B/R has worse perceptual quality. Overall, the best result is attained by the Diffusion-L model, which receives an FSD score of 4.5; this is considerably better than the scores attained by the regression and pix2pix models, which are

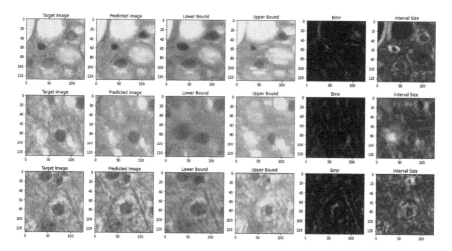

Fig. 3. Examples of per-pixel credible interval bound estimation using generative sampling. The 5th and 95th percentile for each pixel is used as the lower and upper bounds of the credible interval.

624.8 and 54.1, respectively. This perceptual advantage is demonstrated qualitatively in Fig. 2: images generated by Diffusion-L are closer to the target images in both texture and color than pix2pix and the regression model.

It has been theoretically established that attaining a better perceptual score leads to worse performance on distortion [5]. It is thus not surprising that the regression model attains the best distortion measures, as its loss is completely focused on the distortion; as a consequence, its FSD is very poor. Both pix2pix and the diffusion models aim at optimizing a combination of distortion and perceptual measures. Comparing the Diffusion-L and pix2pix models, we note that they have comparable distortion scores, despite the Diffusion-L model's significant performance advantage on perceptual scores.

4.3   Uncertainty Estimation

Figure 3 shows examples of our per-pixel uncertainty estimation. The interval size is the difference between the lower bound and the upper bound of the credible interval, thus larger intervals indicate greater uncertainty. Using our validation set, we observe a calibration factor $\lambda = 1.32$. As we can see in Fig. 3, nuclei are an important source of uncertainty in stains. This finding might motivate the development of future methods which focus on nuclei, e.g. through the use of manual annotation of some nuclei and weighted losses emphasizing these regions.

5   Conclusion

In this work, we demonstrate conditional diffusion models for synthesizing highly realistic histopathology images. We test the perceptual quality of these models using a custom Frechet distance measure. The lack of resolution of the standard Frechet distance FID and the increased discrimination using our custom Frechet distance FSD, indicates that embeddings trained on natural image datasets are not general enough to capture perceptual quality for pathology images. More work is needed to determine whether new quality measures can generalize across a variety of medical image type or must being tailored to each specific image type such as the measure for NASH pathology images in this work. Our results suggest that conditional diffusion models are a promising approach for image-to-image translation tasks, even when we expect outputs with high fidelity and low sample diversity. The observed sample diversity itself can be usefully employed to compute an empirical measure of uncertainty.

Acknowledgements. We thank Shek Azizi, Saurabh Saxena and David Fleet for helpful discussions and suggestions to improve diffusion models. We also thank Yang Wang, Jessica Loo and Peter Cimermancic for their expertise and assistance with the clinical dataset, data processing and digital pathology.

References

1. Angelopoulos, A.N., et al.: Image-to-image regression with distribution-free uncertainty quantification and applications in imaging. In: International Conference on Machine Learning, pp. 717–730. PMLR (2022)
2. Arjovsky, M., Chintala, S., Bottou, L.: Wasserstein generative adversarial networks. In: International Conference on Machine Learning, pp. 214–223. PMLR (2017)
3. Bai, B., Yang, X., Li, Y., Zhang, Y., Pillar, N., Ozcan, A.: Deep learning-enabled virtual histological staining of biological samples (2023). https://doi.org/10.1038/s41377-023-01104-7. https://www.nature.com/articles/s41377-023-01104-7
4. Bates, S., Angelopoulos, A., Lei, L., Malik, J., Jordan, M.: Distribution-free, risk-controlling prediction sets. J. ACM (JACM) **68**(6), 1–34 (2021)
5. Blau, Y., Michaeli, T.: The perception-distortion tradeoff. In: 2018 IEEE/CVF Conference on Computer Vision and Pattern Recognition. IEEE, June 2018. https://doi.org/10.1109/cvpr.2018.00652
6. Dhariwal, P., Nichol, A.: Diffusion models beat gans on image synthesis. Adv. Neural. Inf. Process. Syst. **34**, 8780–8794 (2021)
7. Heusel, M., Ramsauer, H., Unterthiner, T., Nessler, B., Hochreiter, S.: Gans trained by a two time-scale update rule converge to a local nash equilibrium. Advances in neural information processing systems 30 (2017)
8. Ho, J., Jain, A., Abbeel, P.: Denoising diffusion probabilistic models. Adv. Neural. Inf. Process. Syst. **33**, 6840–6851 (2020)
9. Ho, J., Saharia, C., Chan, W., Fleet, D.J., Norouzi, M., Salimans, T.: Cascaded diffusion models for high fidelity image generation. J. Mach. Learn. Res. **23**(47), 1–33 (2022)
10. Horwitz, E., Hoshen, Y.: Conffusion: Confidence intervals for diffusion models (2022). https://doi.org/10.48550/ARXIV.2211.09795. https://arxiv.org/abs/2211.09795
11. Isola, P., Zhu, J.Y., Zhou, T., Efros, A.A.: Image-to-image translation with conditional adversarial networks. In: Proceedings of the IEEE Conference on Computer Vision and Pattern Recognition, pp. 1125–1134 (2017)

12. Kilgour, K., Zuluaga, M., Roblek, D., Sharifi, M.: Fréchet audio distance: a reference-free metric for evaluating music enhancement algorithms. In: INTER-SPEECH, pp. 2350–2354 (2019)
13. Koenker, R., Bassett Jr, G.: Regression quantiles. Econometrica: journal of the Econometric Society, pp. 33–50 (1978)
14. Nichol, A.Q., Dhariwal, P.: Improved denoising diffusion probabilistic models. In: International Conference on Machine Learning, pp. 8162–8171. PMLR (2021)
15. Picon, A., et al.: Autofluorescence image reconstruction and virtual staining for in-vivo optical biopsying. IEEE Access **9**, 32081–32093 (2021). https://doi.org/10.1109/ACCESS.2021.3060926
16. Preuer, K., Renz, P., Unterthiner, T., Hochreiter, S., Klambauer, G.: Fréchet chemnet distance: a metric for generative models for molecules in drug discovery. J. Chem. Inf. Model. **58**(9), 1736–1741 (2018)
17. Ribeiro, M.T., Singh, S., Guestrin, C.: "Why should i trust you?" explaining the predictions of any classifier. In: Proceedings of the 22nd ACM SIGKDD International Conference on Knowledge Discovery and Data Mining, pp. 1135–1144 (2016)
18. Rivenson, Y., Liu, T., Wei, Z., Zhang, Y., de Haan, K., Ozcan, A.: Phasestain: the digital staining of label-free quantitative phase microscopy images using deep learning. Light: Science and Applications (2019). https://doi.org/10.1038/s41377-019-0129-y. https://doi.org/10.1038/s41377-019-0129-y
19. Rivenson, Y., et al.: Virtual histological staining of unlabelled tissue-autofluorescence images via deep learning. Nature Biomed. Eng. **3**(6), 466–477 (2019). https://doi.org/10.1038/s41551-019-0362-y
20. Ronneberger, O., Fischer, P., Brox, T.: U-Net: convolutional networks for biomedical image segmentation. In: Navab, N., Hornegger, J., Wells, W.M., Frangi, A.F. (eds.) MICCAI 2015. LNCS, vol. 9351, pp. 234–241. Springer, Cham (2015). https://doi.org/10.1007/978-3-319-24574-4_28
21. Saharia, C., et al.: Palette: Image-to-image diffusion models. In: ACM SIGGRAPH 2022 Conference Proceedings, pp. 1–10 (2022)
22. Sohl-Dickstein, J., Weiss, E., Maheswaranathan, N., Ganguli, S.: Deep unsupervised learning using nonequilibrium thermodynamics. In: International Conference on Machine Learning, pp. 2256–2265. PMLR (2015)
23. Song, Y., Sohl-Dickstein, J., Kingma, D.P., Kumar, A., Ermon, S., Poole, B.: Score-based generative modeling through stochastic differential equations. arXiv preprint arXiv:2011.13456 (2020)
24. Szegedy, C., Vanhoucke, V., Ioffe, S., Shlens, J., Wojna, Z.: Rethinking the inception architecture for computer vision. In: Proceedings of the IEEE Conference on Computer Vision and Pattern Recognition, pp. 2818–2826 (2016)
25. Tashiro, Y., Song, J., Song, Y., Ermon, S.: Csdi: conditional score-based diffusion models for probabilistic time series imputation. Adv. Neural. Inf. Process. Syst. **34**, 24804–24816 (2021)
26. Zhu, J.Y., Park, T., Isola, P., Efros, A.A.: Unpaired image-to-image translation using cycle-consistent adversarial networks. In: Proceedings of the IEEE International Conference on Computer Vision, pp. 2223–2232 (2017)

Shape-Guided Conditional Latent Diffusion Models for Synthesising Brain Vasculature

Yash Deo[1], Haoran Dou[1], Nishant Ravikumar[1,2], Alejandro F. Frangi[1,2,3,4,5], and Toni Lassila[1,2(✉)]

[1] Centre for Computational Imaging and Simulation Technologies in Biomedicine (CISTIB), School of Computing and School of Medicine, University of Leeds, Leeds, UK

[2] NIHR Leeds Biomedical Research Centre (BRC), Leeds, UK
t.lassila@leeds.ac.uk

[3] Alan Turing Institute, London, UK

[4] Medical Imaging Research Center (MIRC), Electrical Engineering and Cardiovascular Sciences Departments, KU Leuven, Leuven, Belgium

[5] Division of Informatics, Imaging and Data Science, Schools of Computer Science and Health Sciences, University of Manchester, Manchester, UK

Abstract. The Circle of Willis (CoW) is the part of cerebral vasculature responsible for delivering blood to the brain. Understanding the diverse anatomical variations and configurations of the CoW is paramount to advance research on cerebrovascular diseases and refine clinical interventions. However, comprehensive investigation of less prevalent CoW variations remains challenging because of the dominance of a few commonly occurring configurations. We propose a novel generative approach utilising a conditional latent diffusion model with shape and anatomical guidance to generate realistic 3D CoW segmentations, including different phenotypical variations. Our conditional latent diffusion model incorporates shape guidance to better preserve vessel continuity and demonstrates superior performance when compared to alternative generative models, including conditional variants of 3D GAN and 3D VAE. We observed that our model generated CoW variants that are more realistic and demonstrate higher visual fidelity than competing approaches with an FID score 53% better than the best-performing GAN-based model.

Keywords: Image Synthesis · Deep Learning · Brain Vasculature · Vessel Synthesis · Diffusion · Latent Diffusion

1 Introduction

The Circle of Willis (CoW) comprises a complex network of cerebral arteries that plays a critical role in the supply of blood to the brain. The constituent arteries and their branches provide a redundant route for blood flow in the event of occlusion or stenosis of the major vessels, ensuring continuous cerebral perfusion and

A. Mukhopadhyay et al. (Eds.): DGM4MICCAI 2023, LNCS 14533, pp. 164–173, 2024.
https://doi.org/10.1007/978-3-031-53767-7_16

mitigating the risk of ischaemic events [16]. However, the structure of the CoW is not consistent between individuals and dozens of anatomical variants exist in the general population [6,17]. Understanding the differences between these variants is essential to study cerebrovascular diseases, predict disease progression, and improve clinical interventions. Previous studies have attempted to classify and describe the anatomical variations of CoW using categorisations such as the Lippert and Pabst system [6,17]. However, more than 80% of the general population has one of the three most common CoW configurations [2]. The study of anatomical heterogeneity in CoW is limited by the size of available angiographic research data sets, which may only contain a handful of examples of all but the most common phenotypes. The goal of this study is to develop a generative model for CoW segmentations conditioned on anatomical phenotype. Such a model could be used to generate large anatomically realistic virtual cohorts of brain vasculature, and the less common CoW phenotypes can be augmented and explored in greater numbers. Synthesised virtual cohorts of brain vasculature may subsequently be used for training deep learning algorithms on related tasks (e.g. segmenting brain vasculature, classification of CoW phenotype, etc.), or performing in-silico trials.

Generative adversarial networks (GANs) [4] and other generative models have demonstrated success in the synthesis of medical images, including the synthesis of blood vessels and other anatomical structures. However, to the best of our knowledge, no previous study has explored these generative models for synthesising different CoW configurations. Additionally, no previous study has explored the controllable synthesis of different CoW configurations conditioned on desired phenotypes. The synthesis of narrow tubular structures such as blood vessels using conventional generative models is a challenge. Our study builds upon the foundations of generative models in medical imaging and focusses on utilising a conditional latent diffusion model to generate visually realistic CoW configurations with controlled anatomical variations (i.e., by conditioning relevant anatomical information such as CoW phenotypes). Medical images like brain magnetic resonance angiograms (MRA's) tend to be high-dimensional and as a result are prohibitively memory intensive for generative models. Diffusion models and latent diffusion models (LDM) have recently been used for medical image generation [11] and have been shown to outperform GANs in medical image synthesis [18]. Diffusion models have also been successfully used to generate synthetic MRIs [9,19,20] but to the best of our knowledge there are no studies that use latent diffusion models are diffusion models to generate synthetic brain vasculature.

We propose a conditional latent diffusion model that learns latent embeddings of brain vasculature and, during inference, samples from the learnt latent space to synthesise realistic brain vasculature. We incorporate class, shape, and anatomical guidance as conditioning factors in our latent diffusion model, allowing the vessels to retain their shape and allowing precise control over the generated CoW variations. The diffusion model is conditioned to generate different anatomical variants of the posterior cerebral circulation. We evaluate the performance of our model using quantitative metrics such as multiscale structural similarity index (MS-SSIM) and Fr'echet inception distance (FID). Comparative analyses are conducted against alternative generative architectures, including a 3D GAN and a 3D variational auto-encoder (VAE), to assess the superiority of our proposed method in reproducing CoW variations.

2   Methodology

Data and Pre-processing. We trained our model on the publicly available IXI dataset [8] using the 181 3T MRA scans acquired at the Hammersmith Hospital, London. Images were centred, cropped from $512 \times 512 \times 100$ to $256 \times 256 \times 100$, and the intensity normalised. We then used a Residual U-net [10] to extract vessel segmentations from the MRA. The authors manually labelled each case with the presence/absence of one or both peripheral communicating arteries in the CoW. Class 1 includes cases where both the peripheral communication arteries are present (PComA), Class 2 includes cases with only one PComA, while Class 3 includes cases where both PComAs are absent.

Latent Diffusion Model. Recent advances in diffusion models for medical image generation have achieved remarkable success. Diffusion models define a Markov chain of diffusion steps to add random Gaussian noise to the observed data sequentially and then learn to reverse the diffusion process to construct new samples from the noise. Although effective, vanilla diffusion models can be computationally expensive when the input data is of high dimensionality in image space ($256 \times 256 \times 100$ in our study). Hence, we employ the latent diffusion model (LDM), comprising a pretrained autoencoder and a diffusion model. The autoencoder learns a lower-dimensional latent embedding of the brain vasculature, while the diffusion model focusses on modelling the high-level semantic representations in the latent space efficiently.

Following [18], the diffusion process can be defined as forward and reverse Markov chains, where the forward process iteratively transforms the data x_0 (i.e. the latent features from the autoencoder in our approach) into a standard Gaussian X_T as following:

$$q\left(\mathbf{x}_{1:T}|\mathbf{x}_0\right) = \prod_{t=1}^{T} q\left(\mathbf{x}_t|\mathbf{x}_{t-1}\right), q\left(\mathbf{x}_t|\mathbf{x}_{t-1}\right) := \mathcal{N}\left(\mathbf{x}_t; \sqrt{1-\beta_t}\mathbf{x}_{t-1}, \beta_t \mathbf{I}\right)$$

where $q\left(\mathbf{x}_t|\mathbf{x}_{t-1}\right)$ is the transition probability at the time step t based on the noise schedule β_t. Therefore, the noisy data $\mathbf{x}_t$ can be formulated as $q\left(\mathbf{x}_t|\mathbf{x}_0\right) = \mathcal{N}\left(\mathbf{x}_t; \sqrt{\bar{\alpha}_t}\mathbf{x}_0, (1-\bar{\alpha}_t)\mathbf{I}\right)$, where $\alpha_t := 1 - \beta_t, \bar{\alpha}_t := \prod_{s=1}^{t} \alpha_s$.

The reverse process, achieved via a deep neural network parameterised by θ, can then be defined as:

$$p_\theta\left(\mathbf{x}_0|\mathbf{x}_T\right) = p\left(\mathbf{x}_T\right) \prod_{t=1}^{T} p_\theta\left(\mathbf{x}_{t-1}|\mathbf{x}_t\right), p_\theta\left(\mathbf{x}_{t-1}|\mathbf{x}_t\right) := \mathcal{N}\left(\mathbf{x}_{t-1}; \mu_\theta\left(\mathbf{x}_t, t\right), \mathbf{\Sigma}_\theta\left(\mathbf{x}_t, t\right)\right)$$

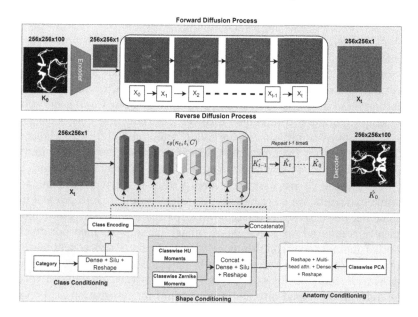

Fig. 1. Overview of the latent diffusion process.

The simplified evidence lower bound (ELBO) loss to optimise the diffusion model by Ho *et al.* [18] can be formulated as a score-matching task where the neural network predicts the actual noise ϵ added to the observed data. The resulting loss function is $\mathcal{L}_\theta := \mathbb{E}_{\mathbf{x}_0, t, C, \epsilon \sim \mathcal{N}(0,1)} \left[\| \epsilon - \epsilon_\theta \left(x_t, t, C \right) \|^2 \right]$ where C is the condition in conditional generation.

We pretrained a multitask attention-based autoencoder using a combination of L1 loss and Dice loss. The encoder transforms the brain image K_0 into a compact latent representation x_0 with dimensions of $256 \times 256 \times 1$. Once the compression model is trained, the latent representations from the training set serve as inputs to the diffusion model for further analysis and generation.

We employ a model with a U-net-based architecture as the diffusion model. Our model has 5 encoding blocks and 5 decoding blocks with skip connections between the corresponding encoding and decoding blocks. We replace the simple convolution layers in the encoding and decoding blocks with a residual block followed by a multihead attention layer to limit information loss in the latent space. Each encoding and decoding block takes the class category (based on CoW phenotypes) as an additional conditional input, while, only the decoding blocks take shape and anatomy features as additional conditional inputs (Fig. 1).

Shape and Anatomy Guidance. Angiographic medical images exhibit intricate anatomical structures, particularly the small vessels in the peripheral cerebral vasculature. Preserving anatomical integrity becomes crucial in the generation of realistic and accurately depicted vessels. However, diffusion models often face challenges in faithfully representing the anatomical structure, which can be attributed to their learning and sampling processes that are heavily based on probability density functions [5]. Previous studies have demonstrated that the inclusion of geometric and shape priors can improve performance in medical image synthesis [1,22]. Additionally, latent space models are susceptible to noise and information loss within the latent space. To this end, we incorporate shape and anatomy guidance to improve the performance of our CoW generation.

The shape guidance component involves incorporating class-wise Hu and Zernike moments as conditions during model training [7,12]. This choice stems from the nature of our image dataset, which comprises both vessel and background regions. By including these shape-related moments as conditions, we aim to better preserve vascular structures within the synthesised images. Hu and Zernike moments are a set of seven invariant moments and a set of orthogonal moments, respectively, commonly used for shape analysis. These moments are typically computed on greyscale or binary images. To incorporate the Hu and Zernike moments as conditions, we first calculate and concatenate these moments for each class. An embedding layer comprising a dense layer with a SiLU activation function [3] and a reshape layer is then introduced to ensure that the data are reshaped into a suitable format for integration as a condition within the decoding branches.

To further enhance the performance of our model, we incorporate anatomy guidance using principal component analysis (PCA) on images from each class. As the majority branches within the CoW exhibit a consistent configuration with minor variations attributed to the presence or absence of specific branches, the model tends to capture an average or mean representation of the CoW and generates synthetic images with very little variation between them. This characteristic becomes significant due to the limited number of images available per class. To address this, we use PCA components as conditions to enable the model to discern distinctive features specific to each class. We extract seven principal components along with the mean component for each class, concatenate them, and reshape the data. The resulting features are then passed through a multi-head attention block, followed by a dense layer and another reshape operation for integration into the decoding branches.

Figure 2 shows the effect of incorporating shape moments and PCA as conditions in our diffusion process. By incorporating shape and anatomy guidance conditions during the training of our diffusion model, we leverage specific features and knowledge related to the vessel structures and the general anatomy of the images. This approach promotes the generation of more realistic images, contributing to an improved anatomical fidelity.

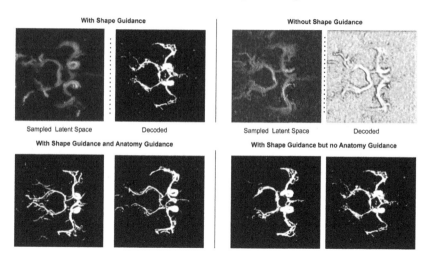

Fig. 2. Row 1: Comparison of output of the latent diffusion network with and without using shape guidance as conditional input. In each column, the image on the left shows the output of our latent diffusion model and the image on the right shows the result of passing the output through the pretrained decoder and obtaining the Maximum Intensity Projection (MIP); Row 2: compares the output of the network with and without using anatomy guidance as conditional input. The generated images displayed on the right, which are produced without the incorporation of anatomy guidance, consistently exhibit a similar variation of the circle of Willis. Conversely, the images presented on the left, which are generated with the inclusion of anatomy guidance, demonstrate a greater degree of realism and variability in the synthesised circle of Willis variations.

3   Experiments and Results

Implementation Details. All models were implemented in TensorFlow 2.8 and Python 3. For the forward diffusion process we use a linear noise schedule with 1000 time steps. The model was trained for 2000 epochs with a learning rate of 0.0005 on a Nvidia Tesla T4 GPU and 38 Gb of RAM with Adam optimiser.

Results and Discussion. To assess the performance of our model, we compared it against two established conditional generative models: 3D C-VAE [13] and a 3D-α-WGAN [14] along with a vanilla LDM and an LDM with shape guidance. We use the FID score to measure the realism of the generated vasculature. To calculate FID we used a pre-trained InceptionV3 as a feature extractor. A lower FID score indicates higher perceptual image quality. In addition, we used MS-SSIM and 4-G-R SSIM to measure the quality of the generated images [15,21]. MS-SSIM and 4-G-R SSIM are commonly used to assess the quality of synthesised images. Typically, a higher score is indicative of better image quality, implying a closer resemblance between the synthesised CoW and the ground truth reference. MS-SSIM and 4-G-R SSIM were calculated over 60 synthesised CoW cases for each model. Table 1 presents the evaluation scores achieved by our model, 3D CVAE, and the 3D-α-WGAN and the above metrics. As seen in Table 1, our model demonstrates a better FID score, suggesting that the distribution of CoW variants synthesised by our model is closer to that observed in real CoW data, compared to the other models. Additionally, our model achieves higher MS-SSIM and 4-G-R SSIM scores compared to the other methods. These higher scores indicate better image quality, implying that the generated CoW samples resemble the real CoW images more closely. Figure 3 provides a qualitative comparison among the generated samples obtained from the three models to provide additional context to the quantitative results presented in Table 1. As the output of each model is a 3D vascular structure, maximum intensity projections (MIP) over the Z-axis which condense the volumetric representation into a 2D plane are used to visually compare the synthesised images.

Table 1. Quantitative evaluation of Synthetic CoW vasculature

Model	FID ↓	MS-SSIM ↑	4-G-R SSIM ↑
3D CVAE	52.78	0.411	0.24
3D-α-WGAN	12.11	0.53	0.41
LDM	176.41	0.22	0.13
LDM + Shape Guidance	8.86	0.58	0.47
Ours (LDM + Shape & Anatomy Guidance)	5.644	0.61	0.51

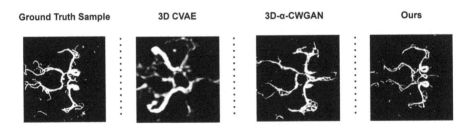

Fig. 3. Comparison between the maximum intensity projections (MIPs) of a real Circle of Willis(CoW) against those synthesised with 3D CVAE, 3D-α-WGAN, and our model.

Figure 3 reveals that the 3D CVAE model can only generate a limited number of major vessels with limited details. On the other hand, although the 3D-α-WGAN model produces the overall structure of the CoW, it exhibits significant anatomical discrepancies with the presence of numerous phantom vessels. On the contrary, our model demonstrates a faithful synthesis of the majority of CoW, with most vessels identifiable. To generate variations of the CoW based on the presence or absence of the posterior communicating artery, our latent diffusion model uses class-conditional inputs where the classes represent different CoW phenotypes. Consequently, to demonstrate the class-conditional fidelity of the proposed approach, we also evaluate the model's performance in a class-wise manner. The qualitative performance of our model for different classes, compared to real images belonging to those classes, is shown in Fig. 3

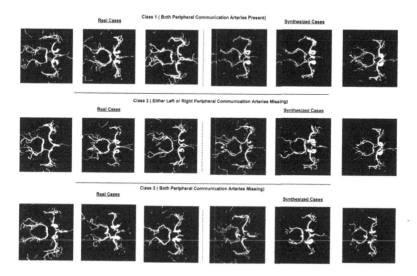

Fig. 4. Comparison between the real and synthesised maximum intensity projections (MIPs) for each of the three classes

Table 2. Quantitative class-wise evaluation of Generated CoW vasculature

Class	FID Score ↓	MS-SSIM ↑	4-G-R SSIM ↑
Class 1	4.41	0.65	0.65
Class 2	3.88	0.52	0.52
Class 3	7.63	0.41	0.41
Overall	5.64	0.61	0.51

The results presented in Fig. 4 demonstrate the performance of our model in generating realistic variations of the Circle of Willis. Particularly notable is the model's proficiency in producing accurate representations for classes 1 and 2, surpassing its performance in class 3 due to the limited sample size of the latter. Our model excels in synthesising the posterior circulation and the middle cerebral arteries, showing remarkable fidelity to anatomical structures. However, it faces challenges in effectively generating continuous representations of the anterior circulation. Further investigation and refinement may be required to enhance the model's ability in this specific aspect. In addition to the visual assessment, we also compute class-wise FID scores, along with the MS-SSIM and 4-G-R SSIM scores. These quantitative evaluations serve to provide a more comprehensive understanding of the model performance with respect to each class. The class-wise performance scores shown in Table 2 are consistent with our observations from Fig. 4, that the model's performance for class 3 is worse than its performance on classes 1 and 2.

4   Conclusion

We proposed a latent diffusion model that used shape and anatomy guidance to generate realistic CoW configurations. Quantitative qualitative results showed that our model outperformed existing generative models based on a conditional 3D GAN and a 3D VAE. Future work will look to enhance the model to capture wider anatomical variability and improve synthetic image quality.

Acknowledgement. This research was partially supported by the National Institute for Health and Care Research (NIHR) Leeds Biomedical Research Centre (BRC) and the Royal Academy of Engineering Chair in Emerging Technologies (CiET1919/19).

References

1. Brooksby, B., Dehghani, H., Pogue, B., Paulsen, K.: Near-infrared (NIR) tomography breast image reconstruction with a priori structural information from MRI: algorithm development for reconstructing heterogeneities. IEEE J. Sel. Top. Quantum Electron. **9**(2), 199–209 (2003)
2. Eftekhar, B., Dadmehr, M., Ansari, S.: Are the distributions of variations of circle of Willis different in different populations? BMC Neurol. **6**(1), 1–9 (2006)
3. Elfwing, S., Uchibe, E., Doya, K.: Sigmoid-weighted linear units for neural network function approximation in reinforcement learning. Neural Netw. **107**, 3–11 (2018)
4. Goodfellow, I., Pouget-Abadie, J., Mirza, M., Xu, B., Warde-Farley, D., Ozair, S., Courville, A., Bengio, Y.: Generative adversarial networks. Commun. ACM **63**(11), 139–144 (2020)
5. Ho, J., Jain, A., Abbeel, P.: Denoising diffusion probabilistic models. Adv. Neural. Inf. Process. Syst. **33**, 6840–6851 (2020)
6. Hoang, T.M., Huynh, T.V., Ly, A.V.H., Pham, M.V.: The variations in the circle of Willis on 64-multislice spiral computed tomography. Trends Med. Sci. **2**(3) (2022)
7. Hu, M.: Visual pattern recognition by moment invariants. IRE Trans. Inf. Theory **8**(2), 179–187 (1962)
8. Information eXtraction from Images Consortium: IXI dataset – brain development. https://brain-development.org/ixi-dataset/. Accessed 14 Feb 2023
9. Jiang, L., Mao, Y., Chen, X., Wang, X., Li, C.: CoLa-Diff: Conditional latent diffusion model for multi-modal MRI synthesis. arXiv preprint arXiv:2303.14081 (2022)

10. Kerfoot, E., Clough, J., Oksuz, I., Lee, J., King, A.P., Schnabel, J.A.: Left-ventricle quantification using residual U-Net. In: Statistical Atlases and Computational Models of the Heart. Atrial Segmentation and LV Quantification Challenges: 9th International Workshop, STACOM 2018, Held in Conjunction with MICCAI 2018, Granada, Spain, September 16, 2018, Revised Selected Papers 9, pp. 371–380. Springer (2019)

11. Khader, F., et al.: Medical diffusion-denoising diffusion probabilistic models for 3D medical image generation. arXiv preprint arXiv:2211.03364 (2022)

12. Khotanzad, A., Hong, Y.: Invariant image recognition by Zernike moments. IEEE Trans. Pattern Anal. Mach. Intell. **12**(5), 199–209 (1990)

13. Kingma, D., Welling, M.: Auto-encoding variational Bayes. arXiv preprint arXiv:1312.6114 (2013)

14. Kwon, G., Han, C., Kim, D.: Generation of 3D brain MRI using auto-encoding generative adversarial networks. Medical Image Computing and Computer Assisted Intervention-MICCAI 2019 22(3), 118–126 (2019)

15. Li, C., Bovik, A.: Content-partitioned structural similarity index for image quality assessment. Signal Processing: Image Communication (2010)

16. Lin, E., Kamel, H., Gupta, A., RoyChoudhury, A., Girgis, P., Glodzik, L.: Incomplete circle of Willis variants and stroke outcome. Eur. J. Radiol. **153**, 110383 (2022)

17. Lippert, H., Pabst, R.: In: Arterial Variations in Man: Classification and Frequency. J.F. Bergmann Verlag, Munich (1985)

18. Müller-Franzes, G., et al.: Diffusion probabilistic models beat GANs on medical images. arXiv preprint arXiv:2212.07501 (2022)

19. Peng, W., Adeli, E., Zhao, Q., Pohl, K.: Generating realistic 3D brain MRIs using a conditional diffusion probabilistic model. arXiv preprint arXiv:2212.08034 (2022)

20. Pinaya, W., et al.: Brain imaging generation with latent diffusion models. In: Deep Generative Models: Second MICCAI Workshop, DGM4MICCAI 2022, Held in Conjunction with MICCAI 2022, pp. 117–126 (2022)

21. Rombach, R., Blattmann, A., Lorenz, D., Esser, P., Ommer: Analyzing the role of visual structure in the recognition of natural image content with multi-scale SSIM. Human Vision and Electronic Imaging XIII, vol. 6806, pp. 410–423. SPIE (2008)

22. Yu, B., Zhou, L., Wang, L., Shi, Y., Fripp, J., Bourgeat, P.: Ea-GANs: edge-aware generative adversarial networks for cross-modality MR image synthesis. IEEE Trans. Med. Imaging **38**(7), 1750–1762 (2019)

Pre-training with Diffusion Models for Dental Radiography Segmentation

Jérémy Rousseau[✉], Christian Alaka, Emma Covili, Hippolyte Mayard, Laura Misrachi, and Willy Au

Allisone Technologies, Paris, France
{jeremy,christian,emma,hippolyte,laura,willy}@allisone.ai
https://www.allisone.ai/

Abstract. Medical radiography segmentation, and specifically dental radiography, is highly limited by the cost of labeling which requires specific expertise and labor-intensive annotations. In this work, we propose a straightforward pre-training method for semantic segmentation leveraging Denoising Diffusion Probabilistic Models (DDPM), which have shown impressive results for generative modeling. Our straightforward approach achieves remarkable performance in terms of label efficiency and does not require architectural modifications between pre-training and downstream tasks. We propose to first pre-train a Unet by exploiting the DDPM training objective, and then fine-tune the resulting model on a segmentation task. Our experimental results on the segmentation of dental radiographs demonstrate that the proposed method is competitive with state-of-the-art pre-training methods.

Keywords: Diffusion · Label-Efficiency · Semantic Segmentation · Dataset Generation

1 Introduction

Accurate automatic semantic segmentation of radiographs is of high interest in the dental field as it has the potential to help practitioners identify anatomical and pathological elements more quickly and precisely. While deep learning methods show robust performances at segmentation tasks, they require a substantial amount of pixel-level annotations which is time-consuming and demands strong expertise in the medical field. Accordingly, many recent state-of-the-art methods [2,5,6,9,22,23] use self-supervised learning as a pre-training step to improve training and reduce labeling effort in computer vision.

C. Alaka, E. Covili, H. Mayard and L. Misrachi—These authors contributed equally to this work.

Supplementary Information The online version contains supplementary material available at https://doi.org/10.1007/978-3-031-53767-7_17.

A. Mukhopadhyay et al. (Eds.): DGM4MICCAI 2023, LNCS 14533, pp. 174–182, 2024.
https://doi.org/10.1007/978-3-031-53767-7_17

Inspired by the renewed interest in denoising for generative modeling, we investigate denoising as a pre-training task for semantic segmentation. Denoising autoencoder is a classic concept in machine learning where a model learns to separate the original data from the noise, and implicitly learns the data distribution by doing so [16,17]. In particular, denoising objective can be easily defined pixel-wise, making it especially well suited for segmentation tasks [4].

Recently, a new class of generative models, known as Denoising Diffusion Probabilistic Models (DDPM) [10,13,15], have shown impressive results for generative modeling. DDPM outperform other state-of-the-art generative models such as Generative Adversarial Networks (GANs) [8] in various tasks, including image synthesis [7].

DDPM learn to convert Gaussian noise to a target distribution via a sequence of iterative denoising steps, yielding impressive results in image synthesis outperforming GANs [7,8].

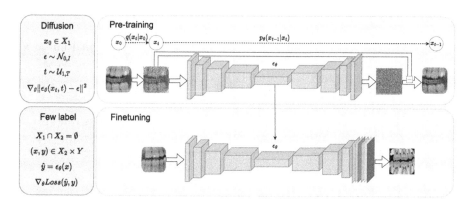

Fig. 1. PTDR method overview. *top* - ϵ_θ is pre-trained on unlabeled dataset X_1 using the training procedure of DDPM [10]. *bottom* - ϵ_θ is then fine-tuned on a small labeled dataset X_2. Y represents the set of ground truth semantic maps.

Following the success of DDPM for generative modeling, [1,18–20] explore their ability to directly generate semantic maps in an iterative process by conditioning each denoising steps with a raw image prior. [3] shows that DDPM are effective representation learners whose feature maps can be used for semantic segmentation, beating previous pre-training methods in a few label regime.

In this paper, we propose Pre-Training with Diffusion models for Dental Radiography segmentation (PTDR). The method consists in pre-training a Unet [14] in a self-supervised manner by exploiting the DDPM training objective, and then fine-tuning the resulting model on a semantic segmentation task.

To sum up our contributions, our method is most similar to [3] but does not require fine-tuning a different model after pre-training. The whole Unet architecture is pre-trained in one step at the difference of [4] which requires two. At inference, only one forward pass is used, making it easier to use than

[1,3]. Finally, we show that our proposed method surpasses other state-of-the-art pre-training methods especially when only few annotated samples are available.

2   Methodology

2.1   Background

Inspired by Langevin dynamics, DDPM [10] formalize the generation task as a denoising problem where an image is gradually corrupted for T steps and then reconstructed through a learned reverse process. Generation is done by applying the reverse process to pure random noise.

Starting from an image $\mathbf{x}_0$, the forward diffusion process iteratively produces noisy versions of the image $\{\mathbf{x_t}\}_{\mathbf{t=1}}^{\mathbf{T}}$, and is defined as a Gaussian Markov chain where $\{\beta_t \in (0,1)\}_{\mathbf{t=1}}^{\mathbf{T}}$ is the variance schedule:

$$q\left(\mathbf{x}_t \mid \mathbf{x}_{t-1}\right) := \mathcal{N}\left(\mathbf{x}_t; \sqrt{1 - \beta_t}\mathbf{x}_{t-1}, \beta_t\mathbf{I}\right) \tag{1}$$

A noisy image $\mathbf{x}_t$ is obtained at any timestep $\mathbf{t}$ from the original image $\mathbf{x}_0$ with the following closed form, let $\alpha_\mathbf{t} = \mathbf{1} - \beta_\mathbf{t}$ and $\bar{\alpha}_t = \prod_{s=1}^{t}\alpha_s$ we have:

$$q\left(\mathbf{x}_t \mid \mathbf{x}_0\right) = \mathcal{N}\left(\mathbf{x}_t; \sqrt{\bar{\alpha}_t}\mathbf{x}_0, (1 - \bar{\alpha}_t)\,\mathbf{I}\right) \tag{2}$$

When the diffusion steps are small enough, the reverse process can also be modeled as a Gaussian Markov chain:

$$p_\theta\left(\mathbf{x}_{t-1} \mid \mathbf{x}_t\right) := \mathcal{N}\left(\mathbf{x}_{t-1}; \mu_\theta\left(\mathbf{x}_t, t\right), \sigma_t^2\mathbf{I}\right) \tag{3}$$

where:

$$\mu_\theta\left(\mathbf{x}_t, t\right) = \frac{1}{\sqrt{\alpha_t}}\left(\mathbf{x}_t - \frac{1 - \alpha_t}{\sqrt{1 - \bar{\alpha}_t}}\epsilon_\theta\left(\mathbf{x}_t, t\right)\right) \quad \sigma_t^2 = \frac{1 - \bar{\alpha}_{t-1}}{1 - \bar{\alpha}_t}\beta_t \tag{4}$$

with ϵ_θ the neural network being optimized.

The training procedure is finally derived by optimizing the usual variational bound on the negative log-likelihood, and consists of randomly drawing samples $\epsilon \sim \mathcal{N}_{0,\mathbf{I}}$, $t \sim \mathcal{U}_{1,\mathbf{T}}$, $\mathbf{x}_0 \sim q(x_0)$ and taking a gradient step on

$$\nabla_\theta\left\|\epsilon_\theta\left(\sqrt{\bar{\alpha}_t}x_0 + \sqrt{1 - \bar{\alpha}_t}\epsilon, t\right) - \epsilon\right\|^2 \tag{5}$$

2.2   DDPM for Semantic Segmentation

The proposed method is based on two steps. First, a denoising model is pre-trained on a large set of unlabeled data following the procedure presented in Sect. 2.1. Second, the model is fine-tuned for semantic segmentation on few annotated data of the same domain by minimizing the cross-entropy loss.

Our method is similar to [3] which leverages a pre-trained DDPM-based model as a feature extractor. Their method involves upsampling feature maps from predetermined activation blocks - from several forward passes at different timesteps - to the target resolution and training an ensemble of pixel-wise classifiers on concatenated feature maps. [3] showed that semantic information carried by feature maps highly depends on the activation block and the diffusion timestep. The latter are thus important hyper-parameters that need to be tuned for each specific semantic task. This method originally introduced in [24] - in the context of GANs - is well-suited for generative models feature extraction but does not leverage the DDPM architecture as PTDR does.

Our approach, by simply re-using the DDPM-trained denoising model for the downstream task, does not need extra classifiers and does not depend on activation blocks hyper-parameter. Moreover, PTDR fine-tuning and inference phases only require one forward pass in which the timestep is fixed to a predetermined value. To that extent, the proposed method is simpler both in terms of training and inference.

3   Experiments and Results

3.1   Experimental Setup

In our experiments, a Unet*[1] based DDPM is trained on unlabeled radiographs, the Unet* is then fine-tuned on a multi-class semantic segmentation task as illustrated in Fig. 1. We experiment with regimes of 1, 2, 5 and 10 training samples and compare our results to other state-of-the-art self-supervised pre-training methods. We used a single NVIDIA T4 GPU for all our experiments.

Datasets: Our main experiment is done on dental bitewing radiographs collected from partner dentists, see Fig. 2. The pre-training dataset contains 2500 unlabeled radiographs. Additionally, 100 bitewing radiographs are fully annotated for 6 classes namely: *dentine, enamel, bone, pulp, other* and *background* as semantic maps; and is randomly split into 10 training, 5 validation, and 85 test samples. There is no intersection between the pre-training and fine-tuning dataset. For our experiments, we use random subsets of the train set of size 1, 2, 5 and 10 respectively. Images are resized to 256×256 resolution and normalized between -1 and 1.

Pre-training: The Unet* implemented in pytorch is trained with a batch size of 2 and follows the training procedure of [7] with 4000 diffusion steps T. We use the official pytorch implementation of [7]. The training was performed for 150k iterations and we saved the weights at iteration 10k, 50k, 100k and 150k for fine-tuning comparison.

Fine-Tuning: The batch size is set at 2. We use a random affine augmentation strategy with the following parameters: rotation angle uniformly sampled from

---

[1] Unet* denotes the specific Unet architecture introduced in [7].

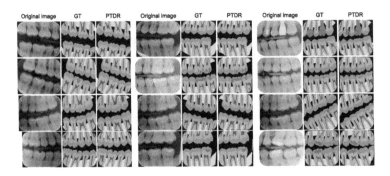

Fig. 2. Comparison on test dental bitewing radiographs of ground truth (GT) against predicted semantic maps from PTDR fine-tuned on 10 labeled images.

$[-180, 180]$, shear sampled from $[-5, 5]$, scale sampled from $[0.9, 1.1]$, and translate factor sampled from $[0.05, 0.05]$. Fine-tuning is done for 200 epochs using the Adam optimizer [11] with a learning rate of $1e^{-4}$, a weight decay of $1e^{-4}$, and a cosine scheduler.

Baseline Methods: The DDPM training procedure is performed for 150k iterations and used for both PTDR and [3] which is referred to as DDPM-MLP for the next sections. We also pre-train a Unet* encoder with MoCo v2 [6] and then fine-tune the whole network on the downstream task. We refer to this method as MoCo v2. Finally, we pre-train a Swin Transformer [12] using SimMIM [22] and use it as an Upernet [21] backbone. We refer to this method as SimMIM. As the Swin backbone relies on batch normalization layers, we do not train SimMIM in the 1-shot regime. For all these methods, we use the same hyper-parameters as proposed in the original papers.

Evaluation Metric: We use mean Intersection over Union (mIoU) as our evaluation metric to measure the performance of the downstream segmentation task.

3.2 Results

We compare our method with other baseline pre-training methods and compare their performances on the multi-class segmentation downstream task in the 10-labeled regime as shown in Table 1.

Our method outperforms all other methods, improving upon the second-best method by 10.5%. Qualitative results on bitewing radiographs are shown in Fig. 3 with predicted semantic maps produced by all compared methods for 1, 5, and 10 training samples. For all regimes, predictions from our method are less coarse than others.

Label Efficiency: In this experiment, we compare our method with baseline methods in different data regimes. Figure 4 illustrates the comparison between methods fine-tuned on 1, 2, 5, and 10 training samples.

Table 1. Comparison of pre-training methods when fine-tuned on 10 labeled samples

Model	Pre-training	mIoU
SwinUperNet	None	59.58
	SimMIM [22]	70.69
Unet*	None	61.40
	MoCo v2 [6]	64.10
	DDPM-MLP [3]	69.64
	PTDR (ours)	**76.96**

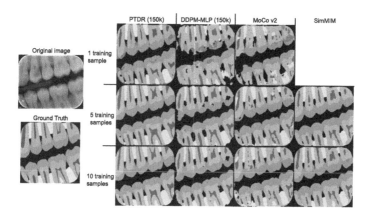

Fig. 3. Semantic maps produced by different methods, PTDR, DDPM-MLP, MoCo v2 and SimMIM. The DDPM pre-training procedure is performed for 150k iterations. Semantic maps were produced by models trained on 1, 5, and 10 training samples to illustrate label efficiency.

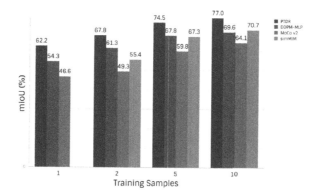

Fig. 4. Label efficiency. Comparison of pre-training methods when fine-tuning in several data regimes (1, 2, 5, and 10 training samples).

Results show that our method yields better performance, in any regime, than all other pre-training methods benchmarked. On average, over all regimes, PTDR improves upon DDPM-MLP, its closest competitor, by 7.08%. Moreover, we can observe in Fig. 4, that our method trained on only 5 training samples outperforms all other methods trained on 10 samples.

Saturation Effect: We explore the influence of the number of DDPM pre-training iterations on the final segmentation performance. In Fig. 5, we observe strong benefits of pre-training between 10k and 50k iterations with an absolute mIoU increase of +7% for PTDR and +6% for DDPM-MLP. As we advance in iteration steps, the pre-training effectiveness decreases. For both methods, we observe that beyond 50k iterations, the performance *saturates* reaching a plateau. This suggests pre-training DDPM can be stopped before reaching ultra-realistic generative performance while still providing an efficient pre-trained model.

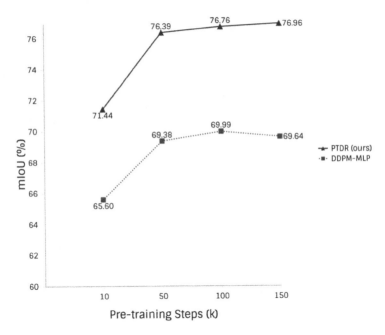

Fig. 5. Saturation effect. Impact of the number of pre-training steps on mIoU for PTDR and DDPM-MLP trained on 10 training samples.

Timestep Influence: We investigate the influence of timestep, which conditions the Unet* and the amount of Gaussian noise added during the diffusion process. We empirically show in Table 2 that timestep 1 is the optimal setup during fine-tuning. This is intuitive as this timestep corresponds to the first diffusion step during which images are almost not corrupted which mirrors the fine-tuning setup on raw images. We did not find any benefits from letting the network learn the timestep value. However, it is worth mentioning that when we do so, the timestep converges to 1.

Table 2. Influence of timestep value on PTDR's fine-tuning performance

Timestep value	1	100	1000	2000	4000	learnt
mIoU	**76.96**	76.94	76.61	74.86	73.60	76.80

Generalization Capacity: In appendix 1, we further investigate the generalization capacity of our method to another medical dataset.

Dataset Generation: In appendix 2, we qualitatively illustrate the method's ability to generate a high-quality artificial dataset with pixel-wise labels.

4 Conclusion

This paper proposes a method that consists of two steps: a self-supervised pre-training using denoising diffusion models training objective and a fine-tuning of the obtained model on a radiograph semantic segmentation task. Experiments on dental bitewing radiographs showed that PTDR outperforms baseline self-supervised pre-training methods in the few label regime. Our simple, yet powerful, method allows the fine-tuning phase to easily exploit all the representations learned in the network during the diffusion pre-training phase without any architectural changes. These results highlight the effectiveness of diffusion models in learning representations. In future works, we will investigate the application of this method to other types of medical datasets.

References

1. Amit, T., Nachmani, E., Shaharbany, T., Wolf, L.: Segdiff: image segmentation with diffusion probabilistic models. arXiv:2112.00390 (2021)
2. Bao, H., Dong, L., Wei, F.: Beit: BERT pre-training of image transformers. arXiv:2106.08254 (2021)
3. Baranchuk, D., Voynov, A., Rubachev, I., Khrulkov, V., Babenko, A.: Label-efficient semantic segmentation with diffusion models. In: International Conference on Learning Representations (2022)
4. Brempong, E.A., Kornblith, S., Chen, T., Parmar, N., Minderer, M., Norouzi, M.: Decoder denoising pretraining for semantic segmentation. Trans. Mach. Learn. Res. (2022)
5. Chen, T., Kornblith, S., Norouzi, M., Hinton, G.: A simple framework for contrastive learning of visual representations. In: International Conference on Machine Learning, pp. 1597–1607. PMLR (2020)
6. Chen, X., Fan, H., Girshick, R., He, K.: Improved baselines with momentum contrastive learning. arXiv:2003.04297 (2020)

7. Dhariwal, P., Nichol, A.: Diffusion models beat GANs on image synthesis. In: Ranzato, M., Beygelzimer, A., Dauphin, Y., Liang, P., Vaughan, J.W. (eds.) Advances in Neural Information Processing Systems, pp. 8780–8794. Curran Associates, Inc. (2021)

8. Goodfellow, I., et al.: Generative adversarial nets. In: Advances in Neural Information Processing Systems (2014)

9. He, K., Fan, H., Wu, Y., Xie, S., Girshick, R.: Momentum contrast for unsupervised visual representation learning. In: Proceedings of the IEEE/CVF Conference on Computer Vision and Pattern Recognition, pp. 9729–9738 (2020)

10. Ho, J., Jain, A., Abbeel, P.: Denoising diffusion probabilistic models. In: Advances in Neural Information Processing Systems, pp. 6840–6851 (2020)

11. Kingma, D.P., Ba, J.: Adam: a method for stochastic optimization. arXiv:1412.6980 (2014)

12. Liu, Z., et al.: Swin transformer: hierarchical vision transformer using shifted windows. In: 2021 IEEE/CVF International Conference on Computer Vision (ICCV) (2021)

13. Nichol, A.Q., Dhariwal, P.: Improved denoising diffusion probabilistic models. In: Proceedings of the 38th International Conference on Machine Learning, ICML 2021, 18–24 July 2021, Virtual Event (2021)

14. Ronneberger, O., Fischer, P., Brox, T.: U-net: convolutional networks for biomedical image segmentation. In: Navab, N., Hornegger, J., Wells, W., Frangi, A. (eds.) MICCAI 2015. LNCS, vol. 9351, pp. 234–241. Springer, Cham (2015). https://doi.org/10.1007/978-3-319-24574-4_28

15. Sohl-Dickstein, J., Weiss, E.A., Maheswaranathan, N., Ganguli, S.: Deep unsupervised learning using nonequilibrium thermodynamics. CoRR (2015)

16. Vincent, P., Larochelle, H., Bengio, Y., Manzagol, P.A.: Extracting and composing robust features with denoising autoencoders. In: Proceedings of the 25th International Conference on Machine Learning. ICML 2008 (2008). https://doi.org/10.1145/1390156.1390294

17. Vincent, P., Larochelle, H., Lajoie, I., Bengio, Y., Manzagol, P.A.: Stacked denoising autoencoders: learning useful representations in a deep network with a local denoising criterion. J. Mach. Learn. Res. (2010)

18. Wolleb, J., Sandkühler, R., Bieder, F., Valmaggia, P., Cattin, P.C.: Diffusion models for implicit image segmentation ensembles. In: International Conference on Medical Imaging with Deep Learning, pp. 1336–1348. PMLR (2022)

19. Wu, J., Fang, H., Zhang, Y., Yang, Y., Xu, Y.: Medsegdiff: medical image segmentation with diffusion probabilistic model. arXiv:2211.00611 (2022)

20. Wu, J., Fu, R., Fang, H., Zhang, Y., Xu, Y.: Medsegdiff-v2: diffusion based medical image segmentation with transformer. arXiv:2301.11798 (2023)

21. Xiao, T., Liu, Y., Zhou, B., Jiang, Y., Sun, J.: Unified perceptual parsing for scene understanding. In: Proceedings of the European Conference on Computer Vision (ECCV) (2018)

22. Xie, Z., et al.: Simmim: a simple framework for masked image modeling. In: Proceedings of the IEEE/CVF Conference on Computer Vision and Pattern Recognition, pp. 9653–9663 (2022)

23. Xu, Z., et al.: Swin MAE: masked autoencoders for small datasets. arXiv:2212.13805 (2022)

24. Zhang, Y., et al.: Datasetgan: efficient labeled data factory with minimal human effort. In: Proceedings of the IEEE/CVF Conference on Computer Vision and Pattern Recognition, pp. 10145–10155 (2021)

ICoNIK: Generating Respiratory-Resolved Abdominal MR Reconstructions Using Neural Implicit Representations in k-Space

Veronika Spieker[1,2(✉)], Wenqi Huang[2], Hannah Eichhorn[1,2], Jonathan Stelter[3], Kilian Weiss[4], Veronika A. Zimmer[1,2], Rickmer F. Braren[3,5], Dimitrios C. Karampinos[3], Kerstin Hammernik[2], and Julia A. Schnabel[1,2,6]

[1] Institute of Machine Learning in Biomedical Imaging, Helmholtz Munich, Neuherberg, Germany
[2] School of Computation, Information and Technology, Technical University of Munich, Munich, Germany
veronika.spieker@helmholtz-munich.de
[3] School of Medicine, Technical University of Munich, Munich, Germany
[4] Philips GmbH, Hamburg, Germany
[5] German Cancer Consortium (DKTK), Partner Site Munich, Munich, Germany
[6] School of Biomedical Engineering and Imaging Sciences, King's College London, London, UK

Abstract. Motion-resolved reconstruction for abdominal magnetic resonance imaging (MRI) remains a challenge due to the trade-off between residual motion blurring caused by discretized motion states and undersampling artefacts. In this work, we generate blurring-free motion-resolved abdominal reconstructions by learning a neural implicit representation directly in k-space (NIK). Using measured sampling points and a data-derived respiratory navigator signal, we train a network to generate continuous signal values. To aid the regularization of sparsely sampled regions, we introduce an additional informed correction layer (ICo), which leverages information from neighboring regions to correct NIK's prediction. The proposed generative reconstruction methods, NIK and ICoNIK, outperform standard motion-resolved reconstruction techniques and provide a promising solution to address motion artefacts in abdominal MRI.

Keywords: MRI Reconstruction · Neural Implicit Representations · Parallel Imaging · Motion-Resolved Abdominal MRI

1 Introduction

Magnetic resonance imaging (MRI) is a non-invasive medical imaging modality with a high diagnostic value. However, its intrinsically long acquisition

Supplementary Information The online version contains supplementary material available at https://doi.org/10.1007/978-3-031-53767-7_18.

A. Mukhopadhyay et al. (Eds.): DGM4MICCAI 2023, LNCS 14533, pp. 183–192, 2024.
https://doi.org/10.1007/978-3-031-53767-7_18

times make MRI more sensitive to motion than other imaging modalities. Especially respiration, which causes local non-rigid deformation of abdominal organs, induces non-negligible motion artefacts. Knowledge of the current state within the breathing cycle, e.g., using external or internal navigators, allows for selection of data to reconstruct from solely one breathing position and reduce artifacts [10]. However, a large portion of data is discarded in this type of reconstruction, resulting in unnecessary prolongation of the acquisition time.

Respiratory-resolved abdominal MRI reconstruction aims to provide high-quality images of one breathing position (typically at end-exhale), while leveraging acquired data points from all states in the respiratory cycle. To effectively utilize information from different breathing states, it is essential to be aware of the specific breathing state during data acquisition. Certain radial sampling trajectories, acquiring data points in a non-Cartesian manner, enable the derivation of a respiratory surrogate signal for motion navigation [17]. Based on such a self-navigator, a common approach is to retrospectively bin the acquired data into multiple motion states and regularize over the motion states to obtain one high-quality reconstruction [2]. While a high number of motion states (i.e., high temporal resolution) decreases residual motion blurring, it minimizes the available data points per motion state. Consequently, undersampling artefacts occur due to the violation of the Nyquist criterion.

Deep learning has emerged as a powerful technique to cope with undersampling in MR, i.e., when fewer data points are available than required to reconstruct an image from the frequency domain [5,7]. Learned denoisers have shown promising results by leveraging information from multiple dynamics [8,9,13], but require pretraining on fully sampled ground truth data. Deep generative models can be trained on acquired undersampled data to infer unavailable information, with the benefit of being independent of such expensive ground truth data [12]. In particular for abdominal motion-resolved MR reconstruction, generative models have been proposed to infer a dynamic sequence of images from a latent space, either directly through a CNN [16] or with intermediate motion modelling [18]. Lately, Feng et al. [1] propose to learn a neural implicit representation of the dynamic image and regularize over multiple dynamics. While these approaches consider data consistency of the predicted images with the original acquired data in k-space, they all rely on binning of the dynamic data to generate images at some point of the training stage, risking residual motion blurring. Additionally, due to the non-uniform sampling pattern, the non-uniform fast Fourier transform (NUFFT) [3], with computationally expensive operations such as regridding and density compensation, is required at each step of the optimization process. Recently, learning a neural implicit representation of k-space (NIK) [6] has shown promising results for binning-free ECG-gated cardiac reconstruction. The training and interpolation is conducted completely in the raw acquisition domain (k-space) and thereby, provide a way to avoid motion binning and expensive NUFFT operations within the optimization and at inference. However, in radial sampling patterns, sampling points sparsify when moving away from the k-space center to high frequency components, resulting in a compromised reconstruction of high frequency information.

In this work, we adapt NIK to respiratory-resolved abdominal MR reconstruction to overcome the challenge of motion-binning of classical reconstruction techniques. We extend NIK's capability to leverage classical parallel imaging concepts, i.e., that missing k-space points can be derived from neighboring points obtained with multiple coils [4]. We perform an informed correction (ICo) by introducing a module which learns this multi-coil neighborhood relationship. Based on the inherently more densely sampled region in the center of k-space, we *inform* the ICo module of the existing relationship by calibrating its weight and use it to *correct* sparsely sampled high frequency regions. Our contributions are three-fold:

1. We modify NIK [6] to learn the first binning-free motion-resolved abdominal reconstruction guided by a data-derived respiratory navigator signal.
2. Inspired by classical MR reconstruction techniques, we perform an informed correction of NIK (ICoNIK), leveraging neighborhood information by applying a kernel which was auto-calibrated on a more densely sampled region.
3. To demonstrate the potential of our work, we present quantitative and qualitative evaluation on retrospectively and prospectively undersampled abdominal MR acquisitions.

2 Methods

2.1 Motion Navigation and Classical Motion-Binned Reconstruction

In 3D abdominal imaging, knowledge of the current motion state can be deducted by acquiring data with a radial stack-of-stars (SoS) trajectory with Cartesian encoding in the feet-head direction. One spoke in the k_x/k_y plane is acquired in each k_z position before moving to the next partition, i.e., next set of spokes (Fig. 1A). By projecting the k-space center of each partition and applying principal component analysis, a 1D curve (nav) indicating the global relative feet-head motion over time t can be derived. Since breathing motion is mainly driven in the feet-head direction, the extracted curve can be used as respiratory navigator signal. We refer the reader to [2] for more details on the respiratory navigator signal derivation.

Consecutively, the navigation signal is used in motion-resolved reconstruction methods to bin the spokes into a pre-defined number of dynamic states n_d (Fig. 1B). A popular representative of motion-resolved reconstruction methods is XD-GRASP [2] which applies a total variation regularization in the dynamic motion state dimension. An inverse Fourier transform can be performed along this dimension and the dynamic images $x = x_{1...n_d}$ are obtained by solving the following optimization problem:

$$\min_x \| \, \mathcal{F}\mathbf{S}x - y \, \|_2^2 + \lambda\mathbf{\Phi}(x). \tag{1}$$

where $y = y_{1...n_d}$ is the multi-coil radial k-space data sorted for each motion state d, $\mathcal{F}$ the NUFFT and $\mathbf{S}$ the coil sensitivity maps. Total variation in the temporal and spatial dimension is imposed by the second term, which consists of the finite difference operator $\mathbf{\Phi}$ and regularization weight $\lambda \in \mathbb{R}^+$.

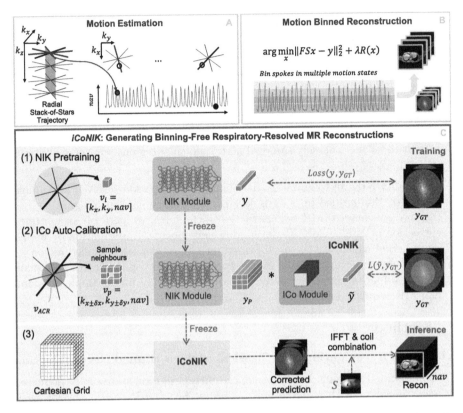

Fig. 1. Graphical overview: (A) Derivation of a motion navigation signal nav from the radial SoS trajectory, associating one nav to each acquired k_x/k_y sample. (B) Classical motion-resolved reconstruction using motion binning based on the surrogate signal. (C) IConIK: (1) NIK is pretrained using samples of one slice obtained with the radial SoS and optimized with the corresponding measured values. Network weights are frozen for further processing. (2) Neighbours of sampling points are queried and their predicted values p fused within the ICo layer. The ICo weights are optimized on a restricted ACR. (3) At inference, a cartesian grid is sampled to generate the k-space and processed to generate dynamic reconstructions.

2.2   Binning-Free Neural Implicit k-Space for Abdominal Imaging

To avoid binning of the acquired data, we propose to learn a continuous implicit representation of k-space (NIK) [6] conditioned on the data-derived respiratory navigator signal. The representation is learned by a neural network G_θ in the form of a mapping from k-space coordinates to the complex signal values. In contrast to the original NIK [6], we predict the signal space for each coil simultaneously rather than including the coil dimension in the input to reduce training effort and enable further post-processing in the coil dimension. Based on the radial sampling trajectory, N coordinates $v_i = [nav_i, k_{xi}, k_{yi}]$ $i = 1, 2, \ldots, N$, $v_i \in \mathbb{R}^3$ representing the current navigator signal value and trajectory position as well as its corresponding coil intensity values $y_i \in \mathbb{R}^{n_c}$ can be queried (Fig. 1C.1). The network G_θ is then optimized with the sampled data pairs (v_i, y_i) to approximate the underlying coordinate-signal mapping:

$$\theta^* = \arg\min_\theta \|G_\theta(v) - y\|_2^2. \tag{2}$$

To account for the increased magnitude in the k-space center, the optimization is solved with a high-dynamic range loss, as described in [6]. At inference, k-space values can be queried for any combination of coordinates within the range of the training data. This allows for sampling based on a Cartesian grid $\bar{v}$ (Fig. 1C), which enables a computationally efficient inverse Fourier transform $\mathcal{F}^{-1}$ instead of NUFFT. Final images are obtained by combining the inverse Fourier-transformed images using the complex conjugate coil sensitivities $\mathbf{S}_c^H$:

$$\hat{x} = \sum_c^{n_c} \mathbf{S}_c^H \mathcal{F}^{-1}(G_{\theta^*}(\bar{v})). \tag{3}$$

2.3   Informed Correction of Neural Implicit k-Space (ICoNIK)

The radial trajectory required for motion navigation and used to train NIK comes at the cost of increased data sparsity towards the outer edges of k-space, which represent the high frequencies and, therefore, details and noise in image domain. To increase the representation capability of NIK, we include neighborhood information by processing NIK's multi-coil prediction with a informed correction module (ICo). Accelerated classical reconstruction methods have shown that k-space values can be derived by linearly combining the neighbouring k-space values of multiple coils. The set of weights can be auto-calibrated on a fully sampled region and consecutively applied to interpolate missing data points [4,11]. We leverage this relationship and correct individual data points with the combination of the surrounding neighbors. As shown in Fig. 1C.2, we sample n_p neighboring points around the input coordinate v_i to obtain $v_p \in \mathbb{R}^{3 \times n_p}$. We predict its multi-coil signal values $y_p \in \mathbb{R}^{n_c \times n_p}$ with NIK and fuse the information in the ICo module using a convolutional network K_ψ to obtain the corrected signal values $\widetilde{y}_i \in \mathbb{R}^{n_c}$ for v_i.

$$\widetilde{y}_i = K_\psi(G_{\theta^*}(v_p)). \tag{4}$$

In general, predictions in the center region of k-space are assumed to be more representative due to the higher amount of ground truth sampling points within this region (inherent by the radial trajectory). Therefore, we "inform" the correction module by calibrating it on NIK's predicted center of k-space, marked as autocalibration region (ACR) in Fig. 1C.2. We select sample points only within a certain distance r to the k-space center for the optimization of the kernel weights ψ^* and at inference, apply the informed correction module to regularize all predictions y_i and output $\widetilde{y}_i$:

$$v_{ACR} \in V(d(v_i) < r), \text{ with } d(v_i) = \sqrt{k_x^2 + k_y^2}, \qquad (5)$$

$$\psi^* = \arg\min_{\psi} \| K_\psi(G_\theta^*(v_{ACR})) - y \|_2^2. \qquad (6)$$

3   Experimental Setup

Data. Free-breathing golden angle stack-of-star acquisitions are obtained at 3T (Ingenia Elition X, Philips Healthcare) with a FOV $= 450 \times 450 \times 252\,\text{mm}^3$, flip angle $= 10°$, voxel size $= 1.5 \times 1.5 \times 3\,\text{mm}^3$, TR/TE1/TE2 $= 4.9/1.4/2.7\,\text{ms}$, $T_{shot} = 395\,\text{ms}$ after approval by the local ethics committee (Klinikum rechts der Isar, 106/20 S-SR). One prolonged sequence and one accelerated were acquired on two separate volunteers, resulting in a set with 1800 (considered as reference R1) and with 600 radial spokes (accelerated by a factor of 3 compared to R1), respectively. Each dataset consists of 84 slices in z dimension and 600 samples per spoke (n_{FE}) were acquired. Due to computational limitations and anisotropic spacing, further processing is conducted on the 2D slices as proof-of-principle. Coil sensitivity maps are estimated using ESPIRiT [14].

Training and Inference. We adapt NIK's architecture (Fourier encoding, 8 layers, 512 features) [6] to output signal values in the coil dimension. We rescale the surrogate motion signal to $[-1,1]$ and use the predefined selected number of spokes (1800 or 600) for training. NIK's training is stopped after 3000 epochs and the model with the lowest residual loss selected for further processing. The ICo module consists of three 3×3 complex kernel layers with interleaved complex ReLUs, and acts on neighboring samples spaced $\delta x, \delta y = n_{FE}^{-1}$ from the original coordinate. The ACR region was empirically determined as r $= 0.4$, and optimization is conducted for 500 epochs. Both modules are trained using an Adam optimizer with a learning rate of $3 \cdot 10^{-5}$ and batch size 10000, optimizing for the linearized high dynamic range loss with $\sigma = 1$, $\epsilon = 1 \cdot 10^{-2}$ and $\lambda = 0.1$. Computations were performed on an NVIDIA RTX A6000, using Python 3.10.1 and PyTorch 1.13.1 (code available at: https://github.com/vjspi/ICoNIK.git). NIK/ICo module training took about 12/4 h and 1.5/0.5 h for the reference and the accelerated version, respectively. Reconstruction of 20 respiratory phases with a matrix size of 300×300 after training takes about 15 s.

Evaluation. Reference motion-resolved 2D slices are reconstructed using inverse NUFFT (INUFFT) and XD-GRASP [2] (total variation in the spatial/temporal domain with factor 0.01/0.1) for 4 motion bins [2]. Results for the end-exhale state (first motion bin) are compared with the adapted NIK [6] and ICoNIK reconstructions at t within the same end-exhale state. For the prolonged sequence, the INUFFT of the gated data from one motion state is considered as approximately fully sampled and used as reference for computing quantitative

Table 1. Quantitative results for 10 slice reconstruction, (*) and (+) mark statistical significance ($p < 0.05$) compared to conventional XD-GRASP and NIK, respectively

	SSIM ↑		PSNR (dB) ↑		NRMSE ↓	
	R1	R3	R1	R3	R1	R3
INUFFT	-	0.80 ± 0.01	-	26.55 ± 0.68	-	0.65 ± 0.03
XD-GRASP	0.85 ± 0.01	0.79 ± 0.01	28.19 ± 0.73	25.73 ± 0.56	0.56 ± 0.03	0.73 ± 0.03
NIK	$0.91 \pm 0.01^*$	$0.82 \pm 0.01^*$	$32.01 \pm 0.89^*$	$27.47 \pm 0.75^*$	$0.35 \pm 0.02^*$	$0.59 \pm 0.03^*$
ICoNIK	$0.91 \pm 0.01^*$	$0.82 \pm 0.01^*$	$32.13 \pm 0.82^*$	$27.36 \pm 0.73^*$	$0.32 \pm 0.02^{*+}$	$0.60 \pm 0.02^*$

evaluation measures, i.e., peak signal-to-noise ratio (PSNR), structural similarity index (SSIM) [15] and normalized root mean squared error (NRMSE) for 10 slices. Qualitative evaluation was performed on the full 1800 spokes (R1), retrospectively and prospectively downsampled 600 spokes (R3-retro/R3-pro).

4   Results

The mean and standard deviation of the reconstruction results from 10 slices compared to the reference are shown in Table 1. Both generative binning-free methods, NIK and ICoNIK, significantly outperform XD-GRASP regarding SSIM, PSNR and NRMSE for both, the reference and accelerated acquisition. ICoNIK additionally shows increased PSNR and significantly reduced NRMSE for the reference scan R1 compared to NIK For R1, a slight decrease of PSNR and increase NRMSE is noticable for ICoNIK compared to NIK.

Qualitative reconstructions are visualized in Fig. 2. Retrospective and prospective undersampling results (R3-retro and R3-pro) show the potential of XD-GRASP, NIK and ICoNIK to leverage information from different motion states to encounter undersampling artefacts originally present in INUFFT. The generative binning-free reconstruction methods (NIK and ICoNIK) indicate sharper images compared to the conventionally binned XD-GRASP, e.g., at the lung-liver edge (green arrow). Furthermore, vessel structures originally visible for the reference (R1-INUFFT) are only represented in NIK and ICoNIK (blue arrow). For the accelerated reconstructions (R3), undersampling artefacts are reduced for all motion-resolved methods and ICoNIK generates smoother reconstruction compared to NIK, while still maintaining structural information. This is supported by the quantitative finding of an increased PSNR for ICoNIK compared to NIK, while maintaining a similar SSIM. Since ICoNIK is capable of interpolating in the time dimension, an arbitrary number of motion states can be generated, allowing for a movie-like reconstruction (see suppl. material).

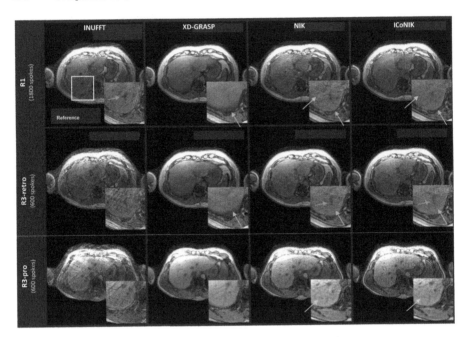

Fig. 2. Motion-resolved reconstructions for the reference (R1), the retrospective down-sampled (R3-retro) and prospectively accelerated (R3-pro) acquisition. Both generative binning-free methods (NIK and ICoNIK) show less motion-blurring compared to traditional motion-binned reconstruction (green arrow) and maintain vessel structures (blue arrow). ICoNIK smooths the reconstruction results compared to NIK (orange arrow). (Color figure online)

5   Discussion and Conclusion

In this work, we showed the potential of learning a neural implicit k-space representation to generate binning-free motion-resolved abdominal reconstructions. We leveraged parallel imaging concepts and induced neighborhood information within an informed correction layer to regularize the reconstruction. Due to the direct optimization in k-space and the learned dynamic representation based on the data-based respiratory navigator, we can generate motion-resolved images at any point of the breathing cycle. Both generative reconstruction methods, NIK and ICoNIK, outperform the traditional reconstruction approach. Additionally, ICoNIK is capable of smoothing reconstructions. While ICoNIK was developed for abdominal MR reconstruction using a radial SoS, we are confident that it can be transferred to other applications and sampling trajectories.

The presented reconstruction technique benefits from its adaptability to non-uniform sampling patterns due to the continuous representation of k-space. Still, the errors may propagate when calibrating ICo on interpolated data, increasing the risk of erroneous calibration weights. The improved performance of ICoNIK for R1 compared to R3, where less ground truth points are available for kernel calibration, supports this indication. Further investigation of the ideal patch retrieval method as well as ACR selection is planned.

ICoNIK is inherently designed to handle inter-subject variability due to subject-specific reconstructions, but requires retraining for each application. Methods for preconditioned neural implicit networks are subject of further development to reduce the retraining cost. Similarly, information transfer between slices could be transferred to facilitate 3D applications at minimized computational cost. Intra-subject variation, e.g., due to changes in breathing pattern, can be captured by ICoNIK as long as the motion navigation signal is representative. Yet, bulk motion and temporal drifts may affect data-based motion surrogates, and thereby influence interpolation capability, motivating us to further look into the robustness of the neural implicit representation.

Lastly, the difficulty to obtain a motion-free ground truth without binning data into motion states remains a significant obstacle in the evaluation process, also for our presented work. More reliable reference acquisition and evaluation techniques are still an active field of research, not only for learning-based reconstruction techniques based on deep generative models. To conclude, our promising findings encourage further investigation of combining traditional parallel imaging concepts with novel deep generative reconstruction models.

Acknowledgements. V.S. and H.E. are partially supported by the Helmholtz Association under the joint research school "Munich School for Data Science - MUDS".

References

1. Feng, J., Feng, R., Wu, Q., Zhang, Z., Zhang, Y., Wei, H.: Spatiotemporal implicit neural representation for unsupervised dynamic MRI reconstruction (2022). arXiv: 2301.00127v2
2. Feng, L., Axel, L., Chandarana, H., Block, K.T., Sodickson, D.K., Otazo, R.: XD-GRASP: golden-angle radial MRI with reconstruction of extra motion-state dimensions using compressed sensing. Magn. Reson. Med. **75**(2), 775–788 (2016)
3. Fessler, J.A.: On NUFFT-based gridding for non-Cartesian MRI. J. Magn. Reson. **188**(2), 191–195 (2007)
4. Griswold, M.A., et al.: Generalized autocalibrating partially parallel acquisitions (GRAPPA). Magn. Reson. Med. **47**(6), 1202–1210 (2002)
5. Hammernik, K., et al.: Learning a variational network for reconstruction of accelerated MRI data. Magn. Reson. Med. **79**(6), 3055–3071 (2018)
6. Huang, W., Li, H.B., Pan, J., Cruz, G., Rueckert, D., Hammernik, K.: Neural implicit k-space for binning-free non-cartesian cardiac MR imaging. In: Frangi, A., de Bruijne, M., Wassermann, D., Navab, N. (eds.) IPMI 2023. LNCS, vol. 13939, pp. 548–560. Springer, Cham (2023). https://doi.org/10.1007/978-3-031-34048-2_42
7. Hyun, C.M., Kim, H.P., Lee, S.M., Lee, S., Seo, J.K.: Deep learning for undersampled MRI reconstruction. Phys. Med. Biol. **63**(13), 135007 (2018)
8. Jafari, R., et al.: GRASPnet: fast spatiotemporal deep learning reconstruction of golden-angle radial data for free-breathing dynamic contrast-enhanced MRI. NMR Biomed. e4861 (2022)

9. Küstner, T., et al.: CINENet: deep learning-based 3D cardiac CINE MRI reconstruction with multi-coil complex-valued 4D spatio-temporal convolutions. Sci. Rep. **10**(1), 13710 (2020)

10. McClelland, J.R., Hawkes, D.J., Schaeffter, T., King, A.P.: Respiratory motion models: a review. Med. Image Anal. **17**(1), 19–42 (2013)

11. Seiberlich, N., Ehses, P., Duerk, J., Gilkeson, R., Griswold, M.: Improved radial GRAPPA calibration for real-time free-breathing cardiac imaging. Magn. Reson. Med. **65**(2), 492–505 (2011)

12. Spieker, V., et al.: Deep learning for retrospective motion correction in MRI: a comprehensive review. IEEE Trans. Med. Imaging **43**(2), 846–859 (2024). https://doi.org/10.1109/TMI.2023.3323215

13. Terpstra, M., Maspero, M., Verhoeff, J., van den Berg, C.: Accelerated respiratory-resolved 4D-MRI with separable spatio-temporal neural networks, arXiv: 2211.05678v1 (2023)

14. Uecker, M., et al.: ESPIRiT-an eigenvalue approach to autocalibrating parallel MRI: where SENSE meets GRAPPA. Magn. Reson. Med. **71**(3), 990–1001 (2014)

15. Wang, Z., Bovik, A.C., Sheikh, H.R., Simoncelli, E.P.: Image quality assessment: from error visibility to structural similarity. IEEE Trans. Image Process. **13**(4), 600–612 (2004)

16. Yoo, J., Jin, K.H., Gupta, H., Yerly, J., Stuber, M., Unser, M.: Time-dependent deep image prior for dynamic MRI. IEEE Trans. Med. Imaging **40**(12), 3337–3348 (2021)

17. Zaitsev, M., Maclaren, J., Herbst, M.: Motion artifacts in MRI: a complex problem with many partial solutions. J. Magn. Reson. Imaging **42**(4), 887–901 (2015)

18. Zou, Q., Torres, L.A., Fain, S.B., Higano, N.S., Bates, A.J., Jacob, M.: Dynamic imaging using motion-compensated smoothness regularization on manifolds (MoCo-SToRM). Phys. Med. Biol. **67**(14) (2022)

Ultrasound Image Reconstruction with Denoising Diffusion Restoration Models

Yuxin Zhang[(✉)], Clément Huneau, Jérôme Idier, and Diana Mateus

Nantes Université, École Centrale Nantes, LS2N, CNRS, UMR 6004,
44000 Nantes, France
yuxin.zhang@ls2n.fr

Abstract. Ultrasound image reconstruction can be approximately cast as a linear inverse problem that has traditionally been solved with penalized optimization using the l_1 or l_2 norm, or wavelet-based terms. However, such regularization functions often struggle to balance the sparsity and the smoothness. A promising alternative is using learned priors to make the prior knowledge closer to reality. In this paper, we rely on learned priors under the framework of Denoising Diffusion Restoration Models (DDRM), initially conceived for restoration tasks with natural images. We propose and test two adaptions of DDRM to ultrasound inverse problem models, DRUS and WDRUS. Our experiments on synthetic and PICMUS data show that from a single plane wave our method can achieve image quality comparable to or better than DAS and state-of-the-art methods. The code is available at https://github.com/Yuxin-Zhang-Jasmine/DRUS-v1/.

Keywords: Ultrasound imaging · Inverse Problems · Diffusion models

1 Introduction

Ultrasound (US) imaging is a popular non-invasive imaging modality, of widespread use in medical diagnostics due to its safety and cost-effectiveness trade-off. Standard commercial scanners rely on simple beamforming algorithms, e.g. Delay-and-Sum (DAS), to transform raw signals into B-mode images, trading spatial resolution for speed. Yet, many applications could benefit from improved resolution and contrast, enabling better organ and lesion boundary detection.

Recent techniques to improve US image quality include adaptive beamforming techniques, e.g. based on Minimum Variance (MV) estimation [2,26], or Fourier-based reconstructions [6]. Other methods focus on optimizing either pre- [1,15] or post-processing steps [16]. Today, there is an increasing interest in model-based approaches [9,20] that better formalize the problem within an

---

Supplementary Information The online version contains supplementary material available at https://doi.org/10.1007/978-3-031-53767-7_19.

A. Mukhopadhyay et al. (Eds.): DGM4MICCAI 2023, LNCS 14533, pp. 193–203, 2024.
https://doi.org/10.1007/978-3-031-53767-7_19

optimization framework. A second branch of methods for improving US image quality leverages the power of Deep Neural Networks (DNNs). Initial approaches in this direction have been trained to predict B-mode images directly [11], the beamforming weights [8,18] or used as post-processing denoisers under supervised training schemes [21,27]. Despite their effectiveness, these methods require datasets of corresponding low-high quality image pairs and therefore do not generalize to other organs/tasks.

Recent hybrid approaches have focused on improving generalizability by combining the best of the model-based and learning worlds. For instance, Chennakeshava et al. [4] propose an unfolding plane-wave compounding method, while Youn et al. [29] combine deep beamforming with an unfolded algorithm for ultrasound localization microscopy. Our work falls within this hybrid model-based deep learning family of approaches [23].

We propose the use of DNN image generators to explore and determine the available solution space for the US image reconstruction problem. In practice, we leverage the recent success of Denoising Diffusion Probabilistic Models (DDPMs) [7,10,19], which are the state-of-the-art in image synthesis in the domain of natural images. More specifically, we build on the Denoising Diffusion Restoration Models (DDRMs) framework proposed by Kawar et al. [14], which adapts DDPMs to various image restoration tasks modeled as linear inverse problems. The main advantage of DDRMs is exploiting the direct problem modeling to bypass the need to retrain DDPMs when addressing new tasks. While the combination of model-based and diffusion models has been explored in the context of CT/MRI imaging [25], this is, to the best of our knowledge, the first probabilistic diffusion model approach for ultrasound image reconstruction.

Our methodological contributions are twofold. First, we adapt DDRMs from restoration tasks in the context of natural images (e.g. denoising, inpainting, superresolution), to the reconstruction of B-mode US images from raw radiofrequency RF channel data. Our approach can be applied to different acquisition types, e.g. sequential imaging, synthetic aperture, and plane-wave, as long as the acquisition can be approximately modeled as a linear inverse problem, i.e. with a model matrix depending only on the geometry and pulse-echo response (point spread function). Our second contribution is introducing a whitening step to cope with the direct US imaging model breaking the *i.i.d.* noise assumption implicit in diffusion models. In addition to the theoretical advances, we provide a qualitative and quantitative evaluation of the proposed approach on synthetic data under different noise levels, showing the feasibility of our approach. Finally, we also demonstrate results on the PICMUS dataset. Next, we review DDRM and introduce our method in Sect. 3.

2   Denoising Diffusion Restoration Models

A DDPM is a parameterized Markov chain trained to generate synthetic images from noise relying on variational inference [7,10,19]. The Markov chain consists of two processes: a forward fixed diffusion process and a backward learned generation process. The forward diffusion process gradually adds Gaussian noise with variance σ_t^2 ($t = 1, \ldots, T$) to the clean signal $\mathbf{x}_0$ until it becomes random noise, while in the backward generation process (see Fig. 1a), the random noise $\mathbf{x}_T$ undergoes a gradual denoising process until a clean $\mathbf{x}_0$ is generated.

(a)

(b)

Fig. 1. (a) DDPMs vs. (b) DDRMs Generation. While DDPMs are unconditional generators, DDRMs condition the generated images to measurements $\mathbf{y}_d$.

An interesting question in model-based deep learning is how to use prior knowledge learned by generative models to solve inverse problems. Denoising Diffusion Restoration Models (DDRM) [14] were recently introduced for solving linear inverse problems, taking advantage of a pre-trained DDPM model as the learned prior. Similar to a DDPM, a DDRM is also a Markov Chain but conditioned on measurements $\mathbf{y}_d$ through a linear observation model $\mathbf{H}_d$ [1]. The linear model serves as a link between an unconditioned image generator and any restoration task. In this way, DDRM makes it possible to exploit pre-trained DDPM models whose weights are assumed to generalize over tasks. In this sense, DDRM is fundamentally different from previous task-specific learning paradigms requiring training with paired datasets. Relying on this principle, the original DDRM paper was shown to work on several natural image restoration tasks such as denoising, inpainting, and colorization.

Different from DDPMs, the Markov chain in DDRM is defined in the spectral space of the degradation operator $\mathbf{H}_d$. To this end, DDRM leverages the Singular Value Decomposition (SVD): $\mathbf{H}_d = \mathbf{U}_d \mathbf{S}_d \mathbf{V}_d^{\mathsf{t}}$ with $\mathbf{S}_d = \text{Diag}(s_1, \ldots, s_N)$, which allows decoupling the dependencies between the measurements. The original observation model $\mathbf{y}_d = \mathbf{H}_d \mathbf{x}_d + \mathbf{n}_d = \mathbf{U}_d \mathbf{S}_d \mathbf{V}_d^{\mathsf{t}} \mathbf{x}_d + \mathbf{n}_d$, can thus be cast as a denoising problem that can be addressed on the transformed measurements:

$$\bar{\mathbf{y}}_d = \bar{\mathbf{x}}_d + \bar{\mathbf{n}}_d$$

with $\bar{\mathbf{y}}_d = \mathbf{S}_d^{\dagger} \mathbf{U}_d^{\mathsf{t}} \mathbf{y}_d$, $\bar{\mathbf{x}}_d = \mathbf{V}_d^{\mathsf{t}} \mathbf{x}_d$, and $\bar{\mathbf{n}}_d = \mathbf{S}_d^{\dagger} \mathbf{U}_d^{\mathsf{t}} \mathbf{n}_d$, where $\mathbf{S}_d^{\dagger}$ is the generalized inverse of $\mathbf{S}_d$. The additive noise $\mathbf{n}_d$ being assumed i.i.d. Gaussian: $\mathbf{n}_d \sim \mathcal{N}(0, \sigma_d^2 \mathbf{I}_N)$, with a known variance σ_d^2 and $\mathbf{I}_N$ the $N \times N$ identity matrix, we then have $\bar{\mathbf{n}}_d$ with standard deviation $\sigma_d \mathbf{S}_d^{\dagger}$.

Each denoising step from $\bar{\mathbf{x}}_t$ to $\bar{\mathbf{x}}_{t-1}$ ($t = T, \ldots, 1$) is a linear combination of $\bar{\mathbf{x}}_t$, the transformed measurements $\bar{\mathbf{y}}_d$, the transformed prediction of $\mathbf{x}_0$ at the current step $\bar{\mathbf{x}}_{\theta,t}$, and random noise. To determine their coefficients which are

---

[1] We use subscript d to refer to the original equations of the DDRM model.

denoted as A, B, C, and D respectively, the condition on the noise, $(A\sigma_t)^2 + (B\sigma_d/s_i)^2 + D^2 = \sigma_{t-1}{}^2$, and on the signal, $A + B + C = 1$, are leveraged, and the two degrees of freedom are taken care of by two hyperparameters.

In this way, the iterative restoration is achieved by the iterative denoising, and the final restored image is $\mathbf{x}_0 = \mathbf{V}_d\overline{\mathbf{x}}_0$. For speeding up this process, skip-sampling [24] is applied in practice. We denote the number of iterations as it.

3   Method: Reconstructing US Images with DDRM

We target the problem of reconstructing US images from raw data towards improving image quality. To model the reconstruction with a linear model, we consider the ultrasonic transmission-reception process under the first-order Born approximation. We introduce the following notations: τ, k, x, and $\mathbf{r}$ respectively denote the time delay, the time index, the reflectivity function, and the observation position in the field of view. When the ultrasonic wave transmitted by the ith element passes through the scattering medium Ω and is received by the jth element, the received echo signal can be expressed as

$$y_{i,j}(k) = \int_{\mathbf{r}\in\Omega} a_i(\mathbf{r})a_j(\mathbf{r})h(k - \tau_{i,j}(\mathbf{r}))x(\mathbf{r})d\mathbf{r} + n_j(k), \qquad (1)$$

where $n_j(k)$ represents the noise for the jth receive element, function h is the convolution of the emitted excitation pulse and the two-way transducer impulse response, and a represents the weights for apodization according to the transducer's limited directivity.

The discretized linear physical model with N observation points and K time samples for all L receivers can then be rewritten as $\mathbf{y} = \mathbf{Hx}+\mathbf{n}$, where $\mathbf{x} \in \mathbb{R}^{N\times1}$, $\mathbf{n} \in \mathbb{R}^{KL\times1}$, and $\mathbf{H} \in \mathbb{R}^{KL\times N}$ is filled with the convolving and multiplying factors from h and a at the delays $\tau_{i,j}$. Due to the Born approximation, the inaccuracy of h and a, and the discretization, the additive noise $\mathbf{n}$ does not only include the white Gaussian electronic noise but also the model error. However, for simplicity, we still assume $\mathbf{n}$ as white Gaussian with standard deviation γ, which is reasonable for the plane wave transmission [5].

While iterative methods exist for solving such linear inverse problems [9,20], our goal is to improve the quality of the reconstructed image by relying on recent advances in diffusion models and, notably, on DDRM. Given the above linear model, we can now rely on DDRM to iteratively guide the reconstruction of the US image from the measurements. However, since DDRM relies on the SVD of $\mathbf{H}$ to go from a generic inverse problem to a denoising/inpainting problem, and since this SVD produces huge orthogonal matrices that cannot be implemented as operators, we propose to transform the linear inverse problem model to:

$$\mathbf{By} = \mathbf{BHx} + \mathbf{Bn}, \qquad (2)$$

where $\mathbf{B} \in \mathbb{R}^{N \times KL}$ is a beamforming matrix that projects channel data to the image domain. After this transformation, we then feed the new inverse problem (Eq. 2) to DDRM to iteratively reconstruct $\mathbf{x}$ from $\mathbf{By}$ observations. In this way, the size of the SVD of $\mathbf{BH}$ becomes more tractable. We call this first model DRUS for Diffusion Reconstruction in US.

However, the noise of the updated direct model $\mathbf{Bn}$ is no longer white and thus, it does not meet the assumption of DDRM. For this reason, we introduce a whitening operator $\mathbf{C} \in \mathbb{R}^{M \times N}$, where $M \leqslant N$, and upgrade the inversion model to its final form:

$$\mathbf{CBy} = \mathbf{CBHx} + \mathbf{CBn}, \tag{3}$$

where $\mathbf{C}$ is such that $\mathbf{CBn}$ is a white noise sequence. In order to compute $\mathbf{C}$, we rely on the eigenvalue decomposition $\mathbf{BB}^{t} = \mathbf{V}\boldsymbol{\Lambda}\mathbf{V}^{t}$ where $\boldsymbol{\Lambda} \in \mathbb{R}^{N \times N}$ is a diagonal matrix of the eigenvalues of $\mathbf{BB}^{t}$, and $\mathbf{V} \in \mathbb{R}^{N \times N}$ is a matrix whose columns are the corresponding right eigenvectors. Then, the covariance matrix of the whitened additive noise $\mathbf{CBn}$ can be written as

$$\mathrm{Cov}(\mathbf{CBn}) = \mathrm{E}[\mathbf{CBnn}^{t}\mathbf{B}^{t}\mathbf{C}^{t}] = \gamma^{2}\mathbf{CBB}^{t}\mathbf{C}^{t} = \gamma^{2}\mathbf{CV}\boldsymbol{\Lambda}\mathbf{V}^{t}\mathbf{C}^{t}.$$

Now, let $\mathbf{C} = \mathbf{P}\boldsymbol{\Lambda}^{-\frac{1}{2}}\mathbf{V}^{t}$ with $\mathbf{P} = [\mathbf{I}_{M}, \mathbf{0}_{M \times (N-M+1)}] \in \mathbb{R}^{M \times N}$. It can be easily checked that $\mathbf{CV}\boldsymbol{\Lambda}\mathbf{V}^{t}\mathbf{C}^{t} = \mathbf{I}_{M}$, proving the noise $\mathbf{CBn}$ is white.

Besides, discarding the smallest eigenvalues by empirically choosing M, rather than strictly limiting ourselves to zero eigenvalues, can compress the size of the observation vector $\mathbf{CBy}$ from $N \times 1$ to $M \times 1$ and make the size of the SVD of $\mathbf{CBH}$ more tractable.

In order to adapt DDRM to the final inverse model in Eq. 3, we consider $\mathbf{y}_d = \mathbf{CBy}$ and $\mathbf{n}_d = \mathbf{CBn}$ as input and compute the SVD of $\mathbf{H}_d = \mathbf{CBH}$. We name this whitened version of the approach WDRUS. In summary:

- DRUS model relies on (2) with $\mathbf{y}_d = \mathbf{By}$, $\mathbf{H}_d = \mathbf{BH}$ and $\mathbf{n}_d = \mathbf{Bn}$
- WDRUS model relies on (3) with $\mathbf{y}_d = \mathbf{CBy}$, $\mathbf{H}_d = \mathbf{CBH}$ and $\mathbf{n}_d = \mathbf{CBn}$.

4   Experimental Validation

In our study, we employed an open-source generative diffusion model [7] at resolution 256×256 pre-trained on ImageNet [22]. We evaluated our method with it $= 50$ on both synthetic data and on the Plane Wave Imaging Challenge in Medical UltraSound (PICMUS) [17] dataset. For the latter, we also experimented with the same unconditional diffusion model but this time fine-tuned with 800 high-quality unpaired ultrasound images acquired with a TPAC Pioneer machine on a CIRS 040GSE phantom. Image samples and data acquisition parameters for fine-tuning are in the supplementary material.

All evaluations in this paper are performed in plane-wave modality. The baseline for synthetic data is beamformed by applying matched filtering $\mathbf{B} = \mathbf{H}^{t}$. The references for the PICMUS dataset apply DAS with 1, 11, and 75 transmissions. Our proposed DRUS and WDRUS are compared against the DAS references by taking measurements of a single-transmission as input.

4.1 Results on Synthetic Data

We simulate data from two phantoms, a synthetic `SynVitro` and Field II `fetus` [12,13]. The model matrix **H** includes receive apodization using Hann window and `f-number = 0.5`, and the beamformer $\mathbf{B} = \mathbf{H}^t$. We simulated channel data $\mathbf{y} = \mathbf{H}\mathbf{x} + \mathbf{n}$ with six levels of additive noise ($\gamma = 0.3, 0.7, 1.0, 1.5, 2.0, 2.5$).

The restoration quality for `SynVitro` is quantitatively evaluated with both resolution and contrast metrics. For the `fetus` phantom, we measure the Structural SIMilarity (SSIM) [28] and the Peak Signal-to-Noise Ratio (PSNR). Resolution is measured as the -6dB Full Width at Half Maximum (FWHM) in axial and lateral directions separately on the six bright scatterers. For evaluating the contrast, we rely on both the Contrast to Noise Ratio (CNR) and the generalized Contrast to Noise Ratio (gCNR):

$$\mathrm{CNR} = 10\log_{10}\left(\frac{|\mu_{\mathrm{in}} - \mu_{\mathrm{out}}|^2}{(\sigma_{\mathrm{in}}^2 + \sigma_{\mathrm{out}}^2)/2}\right), \quad \mathrm{gCNR} = 1 - \int_{-\infty}^{\infty} \min\{f_{\mathrm{in}}(v), f_{\mathrm{out}}(v)\}\, dv,$$

both measured on the four anechoic regions, where the subscripts 'in' and 'out' indicate inside or outside the target regions, v denotes the pixel values, and f refers to the histograms of pixels in each region. The restored images and metrics are summarized in Fig. 2. The metrics for `SynVitro` are averaged over the different noise levels for simplicity.

Qualitatively and quantitatively, both DRUS and WDRUS significantly outperform the matched-filtering baseline $\mathbf{H}^t\mathbf{y}$, and WDRUS is generally superior to

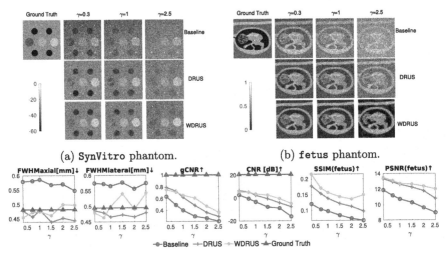

(a) `SynVitro` phantom.   (b) `fetus` phantom.

(c) Comparative evaluation of image quality metrics on synthetic data.

Fig. 2. Comparison of restored images on synthetic data. it = 50 for DRUS and WDRUS. All images are in decibels with a dynamic range [−60,0]. The `fetus` images are normalized between 0 to 1 for calculating the SSIM and PSNR.

DRUS in terms of noise reduction and contrast enhancement. The two proposed approaches even outperform the ground truth for resolution at low-noise conditions (e.g. $\gamma = 0.3, 0.7, 1.0$), while the resolution of images restored by WDRUS under high-noise conditions (e.g. $\gamma = 1.5, 2.0, 2.5$) is worse than that of DRUS.

4.2 Results on PICMUS Dataset

There are four phantoms in the PICMUS [17] dataset. SR and SC are Field II [12, 13] simulations while ER and EC were acquired on a CIRS 040GSE phantom. We use the PICMUS presets where the beamformer **B** comprises receive apodization using Tuckey25 window and f-number $= 1.4$, while **H** has no apodization.

In addition to using FWHM, CNR, and gCNR for evaluating resolution (for SR and ER) and contrast (for SC and EC) introduced in Sect. 4.1, we also use the Signal to Noise Ratio (SNR) μ_{ROI}/σ_{ROI} and the Kolmogorov-Smirnov (KS) test at the 5% significance level, for evaluating the speckle quality preservation (for SC and EC), where ROI is the region of interest. SNR ≈ 1.91 and passing the KS test under a Rayleigh distribution hypothesis are indicators of a good speckle texture preservation. The positions of the ROIs and the p-values of the KS test are in the supplementary material

Using single plane-wave transmission (1PW), we compare our approaches with DAS (1PW, 11PWs, and 75 PWs) qualitatively and quantitatively in Fig. 3 and in Table 1, respectively. We also compare with the scores of six other approaches in Table 1, including the Eigenspace-based Minimum Variance (EMV) [2] which does adaptive beamforming, the traditional Phase Coherence Imaging (PCF) [3], a model-based approach with regularization by denoising (RED) [9], and three learning-based approaches MobileNetV2 (MNV2) [8], Adaptive Ultrasound Beamforming using Deep Learning (ABLE) [18] and DNN-λ^* [30].

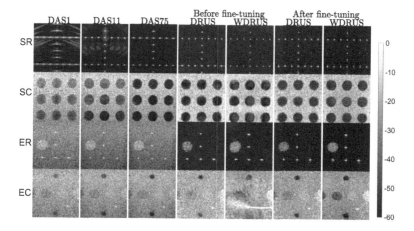

Fig. 3. Reconstructed images comparison on the PICMUS [17] dataset using various approaches. All images are in decibels with a dynamic range $[-60, 0]$.

MNV2 [8] and ABLE [18] are supervised learning techniques designed for adaptive beamforming based on MV estimation. MNV2 [8] utilizes convolution layers to process in-phase and quadrature (IQ) channel data, while ABLE [18] employs fully connected layers to handle radio-frequency (RF) channel data. DNN-λ^* [30] represents a model-based self-supervised approach, relying on a loss function based on prior assumptions of the desired data. It is worth noting that the scores of the first four approaches are sourced from [9], whereas the scores of the latter two approaches are obtained from their respective cited papers.

Table 1. Image quality metrics on the PICMUS SR, SC, ER, EC datasets. A and L denote axial and lateral directions respectively.

	Metric		DAS			no fine-tuning		after fine-tuning		EMV	PCF	RED	MNV2	ABLE	DNN-λ^*
			1	11	75	DRUS	WDRUS	DRUS	WDRUS	[2]	[3]	[9]	[8]	[18]	[30]
SR	FWHM [mm]	A↓	0.38	0.38	0.38	0.30	0.32	0.34	0.31	0.40	0.30	0.37	0.42	0.22	0.28
		L↓	0.81	0.53	0.56	0.47	0.31	0.39	0.28	0.10	0.38	0.46	0.27	0.70	0.32
SC	CNR[dB]↑		10.41	12.86	15.89	16.37	15.20	15.74	16.33	11.21	0.46	15.48	10.48	11.91	10.85
	gCNR↑		0.91	0.97	1.00	0.99	0.99	0.99	0.99	0.93	0.41	0.94	0.89	/	/
	SNR | KS		1.72|✓	1.69|✓	1.68|✓	2.06|✓	1.98|✓	2.03|✓	1.99|✓	/|✓	/|✗	/|✓	/|✓	/	/
ER	FWHM [mm]	A↓	0.56	0.54	0.54	0.34	0.34	0.27	0.22	0.59	5.64	0.48	0.53	/	0.52
		L↓	0.87	0.54	0.56	0.63	1.05	0.55	0.69	0.42	0.76	0.76	0.77	/	0.52
EC	CNR[dB]↑		7.85	11.20	12.00	9.00	-7.25	11.75	13.55	8.10	3.20	14.70	7.80	/	11.6
	gCNR↑		0.87	0.94	0.95	0.88	0.69	0.96	0.97	0.83	0.68	0.98	0.83	/	/
	SNR | KS		1.97|✓	1.91|✓	1.92|✓	1.91|✓	1.50|✗	2.11|✓	1.92|✓	/|✓	/|✗	/|✓	/|✓	/	/

In terms of resolution and contrast, our method is overall significantly better than DAS with 1 plane-wave transmission and can compete with DAS with 75 plane-wave transmissions, as seen in Table 1. However, when the diffusion model is not fine-tuned (using the pre-trained weights from ImageNet [7]), artifacts on the EC image are recovered by WDRUS, which can be explained from two perspectives.

First, while a pre-trained model is a powerful prior and frees the user from acquiring data and training a huge model, there is still a gap between the distribution of natural vs. ultrasound images. This point can be confirmed by comparing the performance of DRUS and WDRUS in Fig. 3 before [col(4,5)] and after [col(6,7)] fine-tuning the diffusion model. With the latter, both DRUS and WDRUS reconstruct images with less distortion, particularly for the anechoic regions on SC and EC, and the hyperechoic region on ER.

Second, due to the approximation of the impulse responses, matrix B has a certain degree of error which is propagated through the eigenvalue decomposition of B and the whitening matrix C. Eventually, such errors may lead to a larger error in WDRUS than in DRUS, as seen when comparing WDRUS [col(5,7)] and DRUS [col(4,6)], despite better contrast and SSIM metrics in Fig. 2c. These errors may also explain why WDRUS is weaker than DRUS in terms of lateral resolution (FWHM L in Table 1) of scatterers in the ER phantom.

Finally, although our method can reconstruct high-quality SR, SC, and EC images using the fine-tuned diffusion model, it is still difficult to retain speckle quality for ER, which is a current limitation.

5   Discussion and Conclusion

Regarding the computing time, our approaches need 3–4 min to form one image, which is slower than DAS1, PCF [3], MNV2 [8], ABLE [18] and DNN-λ^* [30], but faster than EMV [2] and RED [9], which need 8 and 20 min, respectively. RED is slow because each iteration contains an inner iteration while EMV spends time on covariance matrix evaluation and decomposition. Our iterative restoration approaches require multiple multiplication operations with the singular vector matrix, which currently hinders real-time imaging. Accelerating this process is one of our key focuses for future work.

In conclusion, for the first time, we achieve the reconstruction of ultrasound images with two adapted diffusion models, DRUS and WDRUS. Different from previous model-based deep learning methods which are task-specific and require a large amount of data pairs for supervised training, our approach requires none or just a small fine-tuning dataset composed of high-quality (e.g., DAS101) images only (there is no need for paired data). Furthermore, the fine-tuned diffusion model can be used for other US related inverse problems. Finally, our method demonstrated competitive performance compared to DAS75, and other state-of-the-art approaches on the PICMUS dataset.

References

1. Ali, R., Herickhoff, C.D., Hyun, D., Dahl, J.J., Bottenus, N.: Extending retrospective encoding for robust recovery of the multistatic data set. IEEE Trans. Ultrason. Ferroelectr. Freq. Control **67**(5), 943–956 (2020)
2. Asl, B.M., Mahloojifar, A.: Eigenspace-based minimum variance beamforming applied to medical ultrasound imaging. IEEE Trans. Ultrason. Ferroelectr. Freq. Control **57**(11), 2381–2390 (2010)
3. Camacho, J., Parrilla, M., Fritsch, C.: Phase coherence imaging. IEEE Trans. Ultrason. Ferroelectr. Freq. Control **56**(5), 958–974 (2009)
4. Chennakeshava, N., Luijten, B., Drori, O., Mischi, M., Eldar, Y.C., van Sloun, R.J.G.: High resolution plane wave compounding through deep proximal learning. In: IEEE IUS, pp. 1–4 (2020)
5. Chernyakova, T., Cohen, D., Shoham, M., Eldar, Y.C.: iMAP beamforming for high-quality high frame rate imaging. IEEE Trans. Ultrason. Ferroelectr. Freq. Control **66**(12), 1830–1844 (2019)
6. Chernyakova, T., et al.: Fourier-domain beamforming and structure-based reconstruction for plane-wave imaging. IEEE Trans. Ultrason. Ferroelectr. Freq. Control **65**(10), 1810–1821 (2018)
7. Dhariwal, P., Nichol, A.: Diffusion models beat GANs on image synthesis. Technical report, arXiv:2105.05233 (2021)
8. Goudarzi, S., Asif, A., Rivaz, H.: Ultrasound beamforming using MobileNetV2. In: IEEE IUS, pp. 1–4 (2020)

9. Goudarzi, S., Basarab, A., Rivaz, H.: Inverse problem of ultrasound beamforming with denoising-based regularized solutions. IEEE Trans. Ultrason. Ferroelectr. Freq. Control **69**(10), 2906–2916 (2022)
10. Ho, J., Jain, A., Abbeel, P.: Denoising diffusion probabilistic models. Technical report, arXiv:2006.11239 (2020)
11. Hyun, D., Brickson, L.L., Looby, K.T., Dahl, J.J.: Beamforming and speckle reduction using neural networks. IEEE Trans. Ultrason. Ferroelectr. Freq. Control **66**(5), 898–910 (2019)
12. Jensen, J., Svendsen, N.: Calculation of pressure fields from arbitrarily shaped, apodized, and excited ultrasound transducers. IEEE Trans. Ultrason. Ferroelectr. Freq. Control **39**(2), 262–267 (1992)
13. Jensen, J.: Field: a program for simulating ultrasound systems. Med. Biol. Eng. Comput. **34**(Suppl. 1), 351–353 (1997)
14. Kawar, B., Elad, M., Ermon, S., Song, J.: Denoising diffusion restoration models. Technical report, arXiv:2201.11793 (2022)
15. Khan, C., Dei, K., Schlunk, S., Ozgun, K., Byram, B.: A real-time, GPU-based implementation of aperture domain model image reconstruction. IEEE Trans. Ultrason. Ferroelectr. Freq. Control **68**(6), 2101–2116 (2021)
16. Laroche, N., Bourguignon, S., Idier, J., Carcreff, E., Duclos, A.: Fast non-stationary deconvolution of ultrasonic beamformed images for nondestructive testing. IEEE Trans. Comput. Imaging **7**, 935–947 (2021)
17. Liebgott, H., Rodriguez-Molares, A., Cervenansky, F., Jensen, J., Bernard, O.: Plane-wave imaging challenge in medical ultrasound. In: IEEE IUS, pp. 1–4 (2016)
18. Luijten, B., et al.: Adaptive ultrasound beamforming using deep learning. IEEE Trans. Med. Imaging **39**(12), 3967–3978 (2020)
19. Nichol, A., Dhariwal, P.: Improved denoising diffusion probabilistic models. Technical report, arXiv:2102.09672 (2021)
20. Ozkan, E., Vishnevsky, V., Goksel, O.: Inverse problem of ultrasound beamforming with sparsity constraints and regularization. IEEE Trans. Ultrason. Ferroelectr. Freq. Control **65**(3), 356–365 (2018)
21. Perdios, D., Vonlanthen, M., Martinez, F., Arditi, M., Thiran, J.P.: NN-based image reconstruction method for ultrafast ultrasound imaging. IEEE Trans. Ultrason. Ferroelectr. Freq. Control **69**(4), 1154–1168 (2022)
22. Russakovsky, O., et al.: Imagenet large scale visual recognition challenge. Int. J. Comput. Vis. **115**(3), 211–252 (2015)
23. van Sloun, R.J.G., Cohen, R., Eldar, Y.C.: Deep learning in ultrasound imaging. Proc. IEEE **108**(1), 11–29 (2020)
24. Song, J., Meng, C., Ermon, S.: Denoising diffusion implicit models. arXiv:2010.02502 (2020)
25. Song, Y., Shen, L., Xing, L., Ermon, S.: Solving inverse problems in medical imaging with score-based generative models. Technical report, arXiv:2111.08005 (2022)
26. Synnevag, J.F., Austeng, A., Holm, S.: Adaptive beamforming applied to medical ultrasound imaging. IEEE Trans. Ultrason. Ferroelectr. Freq. Control **54**(8), 1606–1613 (2007)
27. Tang, J., Zou, B., Li, C., Feng, S., Peng, H.: Plane-wave image reconstruction via generative adversarial network and attention mechanism. IEEE Trans. Instrum. Meas. **70**, 1–15 (2021)
28. Wang, Z., Bovik, A., Sheikh, H., Simoncelli, E.: Image quality assessment: from error visibility to structural similarity. IEEE Trans. Image Process. **13**(4), 600–612 (2004)

29. Youn, J., Luijten, B., Bo Stuart, M., Eldar, Y.C., van Sloun, R.J.G., Arendt Jensen, J.: Deep learning models for fast ultrasound localization microscopy. In: IEEE IUS, pp. 1–4 (2020)
30. Zhang, J., He, Q., Xiao, Y., Zheng, H., Wang, C., Luo, J.: Ultrasound image reconstruction from plane wave radio-frequency data by self-supervised deep neural network. Med. Image Anal. **70**, 102018 (2021)

Diffusion Model Based Knee Cartilage Segmentation in MRI

Veerasravanthi Mudiyam[1(✉)], Ayantika Das[1], Keerthi Ram[2],
and Mohanasankar Sivaprakasam[1,2]

[1] Indian Institute of Technology, Madras, Chennai, India
ee20s047@smail.iitm.ac.in
[2] Healthcare Technology Innovation Centre, IIT Madras, Chennai, India

Abstract. MRI imaging is crucial for knee joint analysis in osteoarthritis (OA) diagnosis. The segmentation and thickness estimation of knee cartilage are vital steps for OA assessment. Most deep learning algorithms typically produce a single segmentation mask or rely on architectural modifications like Dropout to generate multiple outputs. We propose an alternative approach using Denoising Diffusion Models (DDMs) to yield multiple variants of segmentation outputs for knee cartilage segmentation and thus offer a mechanism to study predictive uncertainty in unseen test data. We further propose to integrate sparsity adaptive losses to supervise the diffusion process to handle intricate knee cartilage structures. We could empirically validate that DDM-based models predict more meaningful uncertainties when compared to Dropout based mechanisms. We have also quantitatively shown that DDM-based multiple segmentation generators are resilient to noise and can generalize to unseen data acquisition setups.

1 Introduction

MRI imaging can capture the structural details of the knee joint highlighting fine morphological changes better than any other imaging modality [2]. The clinical diagnostic protocol for Osteoarthritis (OA) is generally carried out by analyzing MRI scans to delineate the knee cartilages, followed by thickness calculation. The delineation of knee cartilages is often subjective, due to their resemblance to tissue features surrounding the cartilages. When building segmentation algorithms for such structures, having a single annotation restricts the learning, leading to closer mimicking of the available annotation. This also affects the predictive power of segmentation on unseen data.

Deep learning-based algorithms usually result in a single output segmentation, which is typically a single or multi-channel softmax output representing voxel-wise classification posterior probability. If the model has Dropout layers, using them at test-time results in random masking of the layer's inputs, offering an architectural mechanism to obtain variations at the output. We seek to produce an alternative approach for generating multiple segmentation outputs and study it in comparison with the Monte Carlo Dropout technique.

A. Mukhopadhyay et al. (Eds.): DGM4MICCAI 2023, LNCS 14533, pp. 204–213, 2024.
https://doi.org/10.1007/978-3-031-53767-7_20

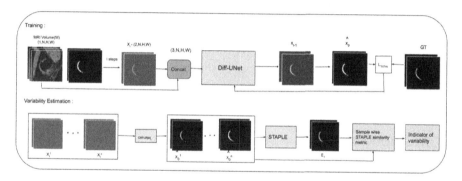

Fig. 1. The outline of our method indicating the integration of the Diff-UNet structure with the losses we have introduced and our devised STAPLE-based mechanism to extract variability generated by the model.

Denoising diffusion models (DDM) [5,10,11] are a new generative method that has emerged as high-quality image generators. They use a learned parametrized iterative denoising process which is the reverse of a Markovian diffusion process to yield a 'sample', and various inverse problems involving image restoration and synthesis have been demonstrated building upon the DDM sampling framework. Specifically, they offer strong sample diversity and faithful mode coverage of the learned data distribution. Both of these are valuable in generalizing segmentation to unseen data under the aleatoric uncertainty of training annotations.

Related Work. U-Net based architectures such as nnUNet [6,8] represent standard baselines in automatic knee cartilage segmentation. Going beyond architectural adaptability, the need for precise segmentation of certain localized and sparse structures led to Attention-based transformer models such as TransUNet [3] which encode strong global context by treating the image features as sequences. Towards supporting application-specific requirements such as thickness measurements, PCAM [8], introduces a morphologically constrained module to ensure continuity in the cartilage segmentation.

DDM-based segmentation models [14–16] can generate multiple samples which are variants of label maps. This is because the input to DDMs is a noised image, and by changing the additive noise, a slightly different sample is yielded at the output. By supervising the diffusion model to generate outputs close to a specified single annotation, we aim to study the characteristics of the generated multiple outputs with regard to two capabilities: *First*, handling noisy MRI scans, *Second*, handling data acquisition variabilities.

For supervising the diffusion process towards segmentation, we have adopted the Diff-UNet [16], and build upon it to address the sparse and intricate characteristics of knee cartilage, which exhibits less inter-tissue variability. Our contributions are:

- a method of leveraging the stochastic capabilities of Diff-UNet to yield multiple variants of segmentation maps for knee cartilage segmentation, offering a mechanism to study predictive uncertainty in unseen test data.
- integration of sparsity adaptive losses to supervise the diffusion process, which has shown quantitative improvement in the segmentation of cartilages in the presence of noise, and for scans acquired from a different setup.

2 Methods

Diffusion UNet. We adopt a new diffusion-based segmentation model Diff-UNet, due to its superior tri-fold capabilities: *First*, Diff-UNet enables volumetric prediction of the segmentation maps, which is essential to capture the complete structure of the cartilages and enforce consistency across multiple 2D slices, which are inherently sparse in appearance. *Second*, Diff-UNet enables multi-label prediction of the segmentation maps, which is vital in labeling the different cartilages which share similar tissue appearances. Diff-UNet enables volumetric multi-label prediction of segmentation maps (x_0) of dimension $N \times W \times H$ by converting it to multi-channel labels (X_0) through one-hot encoding. The iterative noising process generates X_t and $\hat{X}_0$ at each time step, followed by learning to denoise X_t to X_{t-1}, integrating MRI volume $M \in \mathbb{R}^{1 \times N \times W \times H}$ using bi-phased integration: concatenating M with X_t and employing an additional encoder for multi-scale feature maps. The architectural flow of Diff-UNet is represented in Fig. 1.

Third, in Diff-UNet the losses for supervision are enforced on $\hat{X}_0$ predicted at each time step. This is unlike other diffusion models for segmentation which usually do not enforce constraints directly on $\hat{X}_0$, making the Diff-UNet capable of precise structural mapping.

Loss Integration (Diff-UNet$_L$). The enforcement of losses on the predicted $\hat{X}_0$ enables the incorporation of additional losses, which is necessary to better adapt to the sparse knee cartilage structures. We formulate Diff-UNet$_L$ by the addition of boundary enforcement loss (L_{BD}) [7], focal loss (L_{Focal}), and Hausdorff distance-based loss (L_{HD}) [9], along with the existing Diff-UNet losses: MSE loss, Dice loss, and BCE loss $(L_{Diff-UNet})$. Surface losses L_{BD}, L_{HD} are added as both the structures of interest femoral cartilage and tibial cartilage share an adjacent boundary which is difficult for the model to differentiate the dilating boundary. In order to mitigate the challenge posed by the class imbalance problem, we incorporate the L_{Focal} loss, which is designed to tackle the inherent size variation between the tibial, femoral cartilage structures and the non-cartilage regions within the MRI scan. In Fig. 3 the effect of loss integration is indicated by the differences in the segmentation output.

$$L_{\text{total}} = \lambda_1(L_{\text{BD}} + L_{\text{Focal}} + L_{\text{HD}}) + \lambda_2 L_{\text{Diff-UNet}} \quad (1)$$

Multiple Generations and Uncertainty Estimation. The stochastic nature of DDMs enables the generation of multiple segmentation outputs $(\hat{X}_0^i)$ while

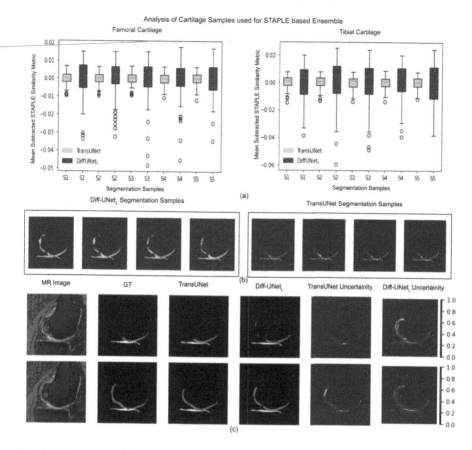

Fig. 2. The top row (a) displays variations in samples from Diff-UNet$_L$ and TransUNet. STAPLE was applied to five samples from each model, and a similarity metric (sensitivity) was calculated between the samples and the STAPLE output. The plots show that TransUNet samples exhibit minimal variation, while Diff-UNet$_L$ samples have a wider spread with some outliers. The second row provides a clearer visualization of the variations, with TransUNet showing concentrated variation and Diff-UNet$_L$ exhibiting meaningful spread. The third row illustrates two consecutive slices with a noticeable abrupt change in GT labels in the right femoral region, where Diff-UNet$_L$ displays more variability in that region.

the deterministic class of models like TransUNet enables stochastic generations can be obtained if the Dropout technique is used. While Dropout based uncertainty stems from the change of configuration of the models, the DDM-based uncertainty highlights the model's uncertainty about the underlying true data distribution. Based on these differences, we aim to investigate the following questions, First, "What are the differences between the samples which are generated from inherent stochastic models and the Dropout simulated ones?", Second, "Are the variations within the samples meaningful and resemble the variations which

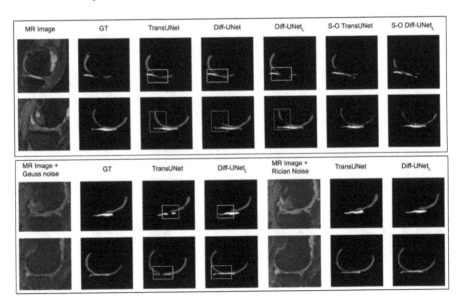

Fig. 3. The top block represents the output of our model compared to the Diff-UNet, TansUNet, and our model, TransUNet in the cross-dataset setup. S-O implies a model trained on the SKM dataset and inferred on the OZ dataset. The second block represents the output of our model compared with TransUNet in two noisy setups. The blue boxes depict better performance of our model, The pink boxes depict better performance of TransUNet. (Color figure online)

can naturally occur during manual annotator based segmentations?" To address these questions, we have proposed the following experimental formulation. We utilized a group of segmentation samples, denoted as $\hat{X}_0^i$, which were processed through the STAPLE [13] algorithm. This allowed us to generate a consensus-based segmentation mask called E_1. This was utilized to measure the similarity between each sample $\hat{X}_0^i$ and E_1, referred to as $STAPLE_{sm}$. A higher degree of similarity between $\hat{X}_0^i$ and E_1 indicates reduced variability among the samples produced by the model. These $STAPLE_{sm}$ was calculated for both diffusion-based model and deterministic segmentation models. We have estimated the Uncertainity from the ensemble of the segmentation samples.

Table 1. Table indicating the performance of our model with the baselines on OZ dataset.

Model	Femoral Cartilage		Tibial Cartilage	
	DSC(%)	ASSD(mm)	DSC(%)	ASSD(mm)
nnUNet	89.03	0.255	86.00	0.211
TransUNet	89.31	0.180	84.82	0.227
nnUNet+PCAM	89.35	0.239	86.11	0.216
Diff-UNet	87.63	0.238	84.44	0.247
Diff-UNet$_L$	88.11	0.210	84.84	0.239

Table 2. Table indicating the performance of the models in different noisy setups on OZ dataset.

Model	Femoral Cartilage (DSC%)			Tibial Cartilage (DSC%)		
	No Noise	Gaussian Noise	Rician Noise	No Noise	Gaussian Noise	Rician Noise
TransUNet	89.31	86.63	87.68	84.82	80.67	82.71
Diff-UNet$_L$	88.11	**86.68**	**87.79**	84.84	**83.59**	**84.26**

Noise Resilience and Generalisation. The segmentation of sparse cartilage structures in knee MRI becomes more challenging when the acquisition is noisy. To assess the noise resilience capability of the Diff-UNet$_L$, we have simulated noisy knee MRI scans by introducing Gaussian noise $\mathcal{N}(\mu, \sigma^2)$ and Rician noise $R(\mu, \sigma^2)$. The adaptability of the models in different acquisition setups is very essential for deploying models in practical use cases. To evaluate the generalizability of Diff-UNet$_L$, we trained it on one dataset and tested its performance on other datasets (cross-dataset setup). This cross-dataset setup poses higher variation within the set due to the OZ dataset being DESS and the SKM dataset being qDESS.

Thickness Estimation. One of the crucial aspects of assessing OA is estimating the thickness of cartilage. In order to better quantify and visualize the segmentation results in terms of clinically relevant metrics, we have adopted a simple yet efficient thickness estimation from [12]. This method creates a refined 3D model, split the mesh into inner and outer components, and computes thickness using the nearest neighbor method. The thickness maps are visualized through 2D projection.

3 Experimental Setup

Datasets. We have made use of two publicly available datasets knee MRI datasets OAI ZIB [1] (OZ) and SKM-Tea dataset [4] (SKM). OZ includes 507 3D DESS MR data with a sagittal acquisition plane with a voxel spacing of $0.3645 \times 0.3645 \times 0.7$ mm. SKM has 155 3D knee MRI volumes acquired using a 5-min 3D quantitative double-echo in steady-state (qDESS) sequence. The voxel spacing is $0.3125 \times 0.3125 \times 0.8$ mm. In order to ensure consistency in OZ data we have adopted the following standardization protocol. We center crop the Region Of Interest (ROI) of the volume with a dimension of $256 \times 256 \times 120$, perform Non-local means filtering, and Normalise intensity levels across volumes. For the SKM dataset such intra-volume variability doesn't exists within a volume, so we have applied only an ROI cropping protocol similar to the OZ dataset. We have considered training and test split as given within the datasets.

Metrics. Dice Similarity Coefficient (DSC), Average symmetric surface Distance (ASSD) are adopted for quantitative analysis between predicted and ground truth. The $STAPLE_{sm}$ is evaluated by calculating the sensitivity between the

Table 3. Table indicating the performance of the models when tested in a cross-dataset setup.

Model	Femoral Cartilage (DSC %)		Tibial Cartilage (DSC %)	
	OZ train SKM test	SKM train OZ test	OZ train SKM test	SKM train OZ test
TransUNet	73.72	72.50	75.66	74.37
Diff-UNet$_L$	**77.20**	**72.64**	**81.51**	**79.19**

samples ($\hat{X}_0^i$) and STAPLE output (E_1). In order to visually highlight the variances of the samples we have considered Mean Subtracted $STAPLE_{sm}$ while plotting as in Fig. 2.

Implementational Details. We have implemented our methods in the PyTorch framework. We assigned higher weightage to sparsity and boundary constraints in the loss function (L_{BD}, L_{Focal}, L_{HD}) where $\lambda_1 = 2$, as compared to $L_{Diff-Unet}$ where $\lambda_2 = 1$. For the model, we have adopted similar parameters as used in the Diff-UNet implementation [16]. We have compared with our performance with nnUNet [6] TransUNet [3], nnUNet+PCAM [8], Diff-UNet [16], Diff-UNet$_L$. For the noisy and generalization case we have compared between TransUNet and Diff-UNet$_L$. The STAPLE-based uncertainty estimation utilized 5 samples per volume. For TransUNet, the Dropout probability is 0.3. We have introduced noise within the volumes by adding Gaussian and Rician noise with $\mathcal{N}(\mu = 0, \sigma^2 = 0.01)$ and $R(\mu = 0, \sigma^2 = 0.01)$ parameters respectively. For the cross-dataset setup, we have trained the models on OZ dataset and inferred on SKM dataset and vice-versa.

Fig. 4. The 2D projection of Thickness maps from ground truth(GT), TransUNet and Diff-UNet$_L$

4   Results

The Fig. 3 qualitatively shows the effect of the additional losses integrated with the Diff-UNet. The addition of losses has ensured better consistency within the femoral and tibial cartilage for Diff-UNet$_L$, as highlighted in the first row of Fig. 3 with blue boxes. From Table 1 we can infer that the results of our model are comparable to the baselines. The mean error thickness values, comparing

GT with respect to Diff-UNet$_L$ and TransUNet femoral cartilage is 0.073 mm & 0.061 mm and tibial cartilage is 0.073 mm & 0.058 mm.

Multiple Segmentation and Uncertainty. From the box plots of Mean subtracted $STAPLE_{sm}$ for TransUNet and Diff-UNet$_L$ in Fig. 2(a), it is clearly quantifiable that the variance of Diff-UNet$_L$ is much higher TransUNet. The median of the box plots of the Diff-UNet$_L$ is higher than that of TransUNet for all the samples. The qualitative visualization of the variations is in Fig. 2(b). In Fig. 2(c), our model effectively detects the uncertain regions in the left femoral regions, which were unmarked by annotators in the first slice but marked in the following slice. This consecutive slice comparison highlights the presence of uncertainty in that specific region. These uncertain regions are well demarcated by our model but missed by TransUNet.

Resilience to Noise. From Table 2, it is clearly evident Diff-UNet$_L$ has better performance than TransUNet in both the noise addition setup. Although in both the cases of Femoral and Tibial cartilage, Diff-UNet$_L$ has better quantification of results, in the latter case the relative increment is much higher when compared to the former. The appearance of the tibial cartilage in MRI scans is more sparse in nature when compared to the Femoral ones, so they have been more affected by the addition of noise. The qualitative visualisation of the results are in the lower block of the Fig. 3.

Generalisation in Cross-Dataset Setup. From Table 3 it is indicative that Diff-UNet$_L$ compared to TransUNet, performs better when the model was trained on OZ dataset & was tested on the SKM dataset and vice versa. Despite the cross-dataset setup, the model has shown incremental performance. The qualitative visualization of the results are in the lower block of the Fig. 3. From the Fig. 4 is observable that the overall structure of the cartilages predicted by the Diff-UNet$_L$ is relatively smooth.

5 Discussions

The integration of losses has shown better performance mostly in predicting the cartilages since they are sparse structures in MRI and need additional enforcement. The better quantification of variances and qualification of uncertainty maps from our model are due to DDM's capability of providing meaningful variations when there is allowable stochasticity. This is further attributed to the model's capacity to generalize beyond specific annotations and adapt to the intrinsic structures present in the scans, despite being trained on a single annotation. Diff-UNet$_L$ outperforms in noisy setups due to DDMs' unique denoising-based sampling process, enabling better adaptation to noisy conditions during mapping from Gaussian to target distributions. The better generalization of the model is due to the fact that DDMs can better capture the distributional properties of the target without being biased to a certain set of data shown to the model during training.

6   Conclusion

Our proposed DDM-based multiple segmentation generator has shown to have a higher variability within the regions of generations which are natural causes of uncertainty while manual annotation. We have quantitatively and qualitatively verified that diffusion-based models better highlight uncertainty than Droput-based techniques. We have shown that after the addition of Gaussian and Rician noise, our model has better DSC % as compared to TransUNet. Also, in the cross-dataset setup, our method has better performance.

References

1. Ambellan, F., Tack, A., Ehlke, M., Zachow, S.: Automated segmentation of knee bone and cartilage combining statistical shape knowledge and convolutional neural networks: data from the osteoarthritis initiative. Med. Image Anal. **52**, 109–118 (2019)
2. Braun, H.J., Gold, G.E.: Diagnosis of osteoarthritis: imaging. Bone **51**(2), 278–288 (2012)
3. Chen, J., et al.: TransUNet: transformers make strong encoders for medical image segmentation. arXiv preprint arXiv:2102.04306 (2021)
4. Desai, A.D., et al.: SKM-TEA: a dataset for accelerated MRI reconstruction with dense image labels for quantitative clinical evaluation. arXiv preprint arXiv:2203.06823 (2022)
5. Ho, J., Jain, A., Abbeel, P.: Denoising diffusion probabilistic models. In: Advances in neural information processing systems, vol. 33, pp. 6840–6851 (2020)
6. Isensee, F., Jaeger, P.F., Kohl, S.A., Petersen, J., Maier-Hein, K.H.: nnU-Net: a self-configuring method for deep learning-based biomedical image segmentation. Nat. Methods **18**(2), 203–211 (2021)
7. Kervadec, H., Bouchtiba, J., Desrosiers, C., Granger, E., Dolz, J., Ayed, I.B.: Boundary loss for highly unbalanced segmentation. In: International Conference on Medical Imaging with Deep Learning, pp. 285–296. PMLR (2019)
8. Liang, D., Liu, J., Wang, K., Luo, G., Wang, W., Li, S.: Position-prior clustering-based self-attention module for knee cartilage segmentation. In: Medical Image Computing and Computer Assisted Intervention-MICCAI 2022: 25th International Conference, Singapore, 18–22 September 2022, Proceedings, Part V, pp. 193–202. Springer, Cham (2022). https://doi.org/10.1007/978-3-031-16443-9_19
9. Ma, J., et al.: How distance transform maps boost segmentation CNNs: an empirical study. In: Arbel, T., Ayed, I.B., de Bruijne, M., Descoteaux, M., Lombaert, H., Pal, C. (eds.) Medical Imaging with Deep Learning. Proceedings of Machine Learning Research, vol. 121, pp. 479–492. PMLR, 06–08 July 2020. http://proceedings.mlr.press/v121/ma20b.html
10. Peng, W., Adeli, E., Zhao, Q., Pohl, K.M.: Generating realistic 3D brain MRIs using a conditional diffusion probabilistic model. arXiv preprint arXiv:2212.08034 (2022)
11. Pinaya, W.H., et al.: Brain imaging generation with latent diffusion models. In: Mukhopadhyay, A., Oksuz, I., Engelhardt, S., Zhu, D., Yuan, Y. (eds.) MICCAI Workshop on Deep Generative Models, vol. 13609, pp. 117–126. Springer, Cham (2022). https://doi.org/10.1007/978-3-031-18576-2_12

12. Sahu, P., et al.: Reproducible workflow for visualization and analysis of osteoarthritis abnormality progression. In: Proceedings of the International Workshop on Quantitative Musculoskeletal Imaging (QMSKI) (2022)
13. Warfield, S.K., Zou, K.H., Wells, W.M.: Simultaneous truth and performance level estimation (staple): an algorithm for the validation of image segmentation. IEEE Trans. Med. Imaging **23**(7), 903–921 (2004)
14. Wolleb, J., Sandkühler, R., Bieder, F., Valmaggia, P., Cattin, P.C.: Diffusion models for implicit image segmentation ensembles. In: International Conference on Medical Imaging with Deep Learning, pp. 1336–1348. PMLR (2022)
15. Wu, J., Fang, H., Zhang, Y., Yang, Y., Xu, Y.: MedSegDiff: medical image segmentation with diffusion probabilistic model. arXiv preprint arXiv:2211.00611 (2022)
16. Xing, Z., Wan, L., Fu, H., Yang, G., Zhu, L.: Diff-UNet: a diffusion embedded network for volumetric segmentation. arXiv preprint arXiv:2303.10326 (2023)

Semantic Image Synthesis for Abdominal CT

Yan Zhuang, Benjamin Hou, Tejas Sudharshan Mathai, Pritam Mukherjee, Boah Kim, and Ronald M. Summers[✉]

Imaging Biomarkers and Computer-Aided Diagnosis Laboratory, Department of Radiology and Imaging Sciences, National Institutes of Health Clinical Center, Bethesda, USA
{yan.zhuang2,benjamin.hou,tejas.mathai,pritam.mukherjee,boah.kim,
rsummers}@nih.gov

Abstract. As a new emerging and promising type of generative models, diffusion models have proven to outperform Generative Adversarial Networks (GANs) in multiple tasks, including image synthesis. In this work, we explore semantic image synthesis for abdominal CT using conditional diffusion models, which can be used for downstream applications such as data augmentation. We systematically evaluated the performance of three diffusion models, as well as to other state-of-the-art GAN-based approaches, and studied the different conditioning scenarios for the semantic mask. Experimental results demonstrated that diffusion models were able to synthesize abdominal CT images with better quality. Additionally, encoding the mask and the input separately is more effective than naïve concatenating.

Keywords: CT · Abdomen · Diffusion model · Semantic Image Synthesis

1 Introduction

Semantic image synthesis aims to generate realistic images from semantic segmentation masks [17]. This field has a broad range of applications that range from data augmentation and anonymization to image editing [4,7,12–14,20]. For instance, Lau *et al.* used a conditional Generative Adversarial Network (GAN) and semantic label maps to synthesize scar tissues in cardiovascular MRI for data augmentation [12]. Hou *et al.* employed a StyleGAN to synthesize pathological retina fundus images from free-hand drawn semantic lesion maps [7]. Shin *et al.* utilized a conditional GAN to generate abnormal MRI images with brain tumors, and to serve as an anonymization tool [20]. Mahapatra *et al.* leveraged a conditional GAN to synthesize chest x-ray images with different disease characteristics by conditioning on lung masks [13]. Blanco *et al.* proposed editing histopathological images by applying a set of arithmetic operations in the GANs' latent space [4]. The main objectives of these works were to address the issues

A. Mukhopadhyay et al. (Eds.): DGM4MICCAI 2023, LNCS 14533, pp. 214–224, 2024.
https://doi.org/10.1007/978-3-031-53767-7_21

of data scarcity, given the time-consuming, labor-intensive, and extremely costly of obtaining high-quality data and annotations [5]. These studies have demonstrated the effectiveness of using synthetic data for downstream tasks, assuming that GAN-based generative models can generate photo-realistic images.

More recently, several studies have illustrated that diffusion models surpassed GAN-based models in multiple image synthesis tasks [2,18], demonstrating an ability to generate realistic and high-fidelity images. Similarly, diffusion models draw an increasing attention in medical imaging, including registration [11], segmentation [25], reconstruction [1], image-to-image translation [16], anomaly detection [24], and *etc*. Kazerouni *et al*. provided a comprehensive review on latest research progress regarding diffusion models for medical imaging [10]. Despite being the *de facto* standard for image synthesis, the application of diffusion models for medical semantic image synthesis remain relatively unexplored. To the best of our knowledge, few studies exist for this task. Zhao *et al*. employed the Semantic Diffusion Model (SDM) to synthesize pulmonary CT images from segmentation maps for data augmentation [27], while Dorjsembe *et al*. developed a diffusion model to simulate brain tumors in MRI [3].

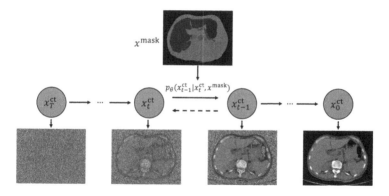

Fig. 1. The semantic image synthesis for abdominal CT using a diffusion model. The CT image mask x^{mask} guides the diffusion process, which synthesizes a CT image that matches the semantic layout of the mask x^{mask}. p_θ is the neural network parameterized by θ, and x_t^{ct} is the output of the network at time step t. Different colors of x^{mask} represent different abdominal organs and structures.

As prior work primarily used GAN-based models and focused on the head and thorax, our study investigates the use of conditional diffusion models for the semantic medical image synthesis of abdominal CT images as shown in Fig. 1. The abdomen is anatomically complex with subtle structures (e.g., lymph nodes) interwoven with large organs (e.g., liver). Consequently, the associated semantic segmentation maps are dense and complex. This complexity presents a far greater challenge when conducting image synthesis for the abdomen. We initially explored different conditioning configurations for diffusion models, such as channel-wise concatenation, encoding the mask in a separated encoder, and the use of auxiliary information. Then we assessed the performance of the conditional diffusion models against GAN-based models in terms of image quality and learned correspondence. Our experimental evaluation demonstrated that encoding the mask and the input enabled the diffusion model to converge earlier and gain improved performance. Moreover, the results showed that conditional diffusion models achieved superior image quality in terms of Fréchet Inception Distance score (FID), Structural Similarity Index Measure (SSIM) and Peak Signal to Noise Ratio (PSNR) scores within a large, publicly available dataset. While diffusion models excelled at learned correspondence in large organs, they were outperformed by GAN-based methods in small structures and organs. Despite this, the conditional diffusion models still yielded promising results.

Our contributions are two-fold: (1) we demonstrate the effectiveness of diffusion models in the task of semantic image synthesis for abdomen CT and provided a comprehensive comparative evaluation against other State-of-The-Art (SOTA) GAN-based approaches; (2) we empirically show that encoding masks in a separated encoder branch can achieve superior performance, shedding light on finding a more effective way to leverage the semantic mask information.

2   Method

Although the process of synthesizing a CT image from a given semantic segmentation mask is a form of conditional image generation, it fundamentally relies on (unconditional) diffusion models. This study focuses on Denoising Diffusion Probabilistic Models (DDPM) [15].

The DDPM model consists of a forward diffusion process and a reverse diffusion process. The forward process progressively transforms a clean image into an image with isotropic Gaussian noise. Mathematically, considering a clean image sample x_0^{ct} and a set of time steps $\{1, \cdots, t, \cdots, T\}$, Gaussian noise is progressively added to the image at time step t by:

$$q(x_t^{ct}|x_{t-1}^{ct}) = \mathcal{N}(x_t^{ct}; \sqrt{1-\beta_t}x_{t-1}^{ct}, \beta_t \mathbf{I}), \tag{1}$$

where β_t is the scheduled variance. Then, using Markov chain rule, the forward process of x_t^{ct} from x_0^{ct} can be formulated by:

$$q(x_t^{ct}|x_0^{ct}) = \mathcal{N}(x_t^{ct}; \sqrt{\bar{\alpha}_t}x_0^{ct}, (1-\bar{\alpha})\mathbf{I}), \tag{2}$$

where $\alpha_t = 1 - \beta_t$ and $\bar{\alpha}_t = \prod_{s=1}^{t} \alpha_s$. Accordingly, for $\epsilon \in \mathcal{N}(0, \mathbf{I})$, a noisy image x_t can be expressed in terms of x_0 in a closed form:

$$x_t^{ct} = \sqrt{\bar{\alpha}_t}x_0^{ct} + \sqrt{1-\bar{\alpha}_t}\epsilon. \tag{3}$$

On the other hand, the reverse diffusion process gradually removes Gaussian noise by approximating $q(x_{t-1}|x_t)$ through a neural network p_θ parameterized by θ,

$$p_\theta(x_{t-1}^{ct}|x_t^{ct}) = \mathcal{N}(x_{t-1}^{ct}; \mu_\theta(x_t^{ct}, t), \Sigma_\theta(x_t^{ct}, t)), \tag{4}$$

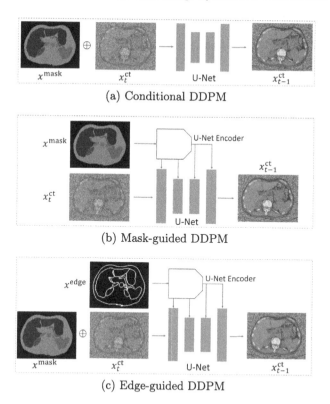

(a) Conditional DDPM

(b) Mask-guided DDPM

(c) Edge-guided DDPM

Fig. 2. Different conditions of diffusion models for semantic image synthesis of abdominal CT images: (a) channel-wise concatenating, denoted as "conditional DDPM"; (b) mask guidance where the conditioning mask is encoded in a U-Net encoder, denoted as "mask-guided DDPM"; (c) using other auxiliary information, e.g., semantic edge map, denoted as "edge-guided DDPM".

where μ_θ and Σ_θ are predicted mean and variance. Thus, the image sample x_{t-1} at time step $t-1$ can be predicted as:

$$x_{t-1}^{ct} = \frac{1}{\sqrt{\alpha_t}} \left(x_t^{ct} - \frac{\beta_t}{\sqrt{1-\bar{\alpha}_t}} \epsilon_\theta(x_t^{ct}, t) \right) + \sigma_t \mathbf{z}, \tag{5}$$

where ϵ_θ is the trained U-Net, σ_t is the learned variance, and $\mathbf{z} \in \mathcal{N}(0, \mathbf{I})$. A detailed formulation of DDPM can be found in [15].

The aforementioned formulation of DDPM is an unconditional image synthesis process, meaning that the synthetic CT images are generated from random anatomic locations. However, our goal is a conditional image synthesis process. The aim is to generate the CT images in such a way that the synthetic CT images preserve the same semantic layout as the given input CT masks. The input CT image mask, denoted by x^{mask}, should guide the diffusion process and synthesize an image that matches the semantic layout of the mask. In this pilot work, to assess the effectiveness of various conditioning methods, we have presented three different conditioning scenarios: (1) channel-wise concatenating, denoted as "conditional DDPM"; (2) mask guidance where the conditioning mask is encoded in a separated network branch, denoted as "mask-guided DDPM"; (3) using other auxiliary information, e.g., semantic edge map, denoted as "edge-guided DDPM".

Conditional DDPM. In this method, the idea was to concatenate the mask x^{mask} together with the input image x_t in an additional input channel. The network architecture was as the same as the DDPM model, as shown in Fig. 2(a). Then, in this case, Eq. (5) became:

$$x_{t-1}^{ct} = \frac{1}{\sqrt{\alpha_t}} \left(x_t^{ct} - \frac{\beta_t}{\sqrt{1 - \bar{\alpha}_t}} \epsilon_\theta(x_t^{ct} \oplus x^{mask}, t) \right) + \sigma_t \mathbf{z}, \qquad (6)$$

where $x_t^{ct} \oplus x^{mask}$ is the channel-wise concatenation of the input image x_t and the given mask x^{mask}.

Mask-Guided DDPM. The second strategy was to encode the mask separately by employing another U-Net-like encoder, and injecting the encoding information directly into the main U-Net branch. More specifically, feature maps from the convolutional layers before each downsampling layer of the U-Net-like encoder were concatenated to the corresponding feature maps of the main U-Net branch encoder and decoder, as shown in Fig. 2(b). This idea was similar to SPADE [17]. Then in this case, Eq. (5) became:

$$x_{t-1}^{ct} = \frac{1}{\sqrt{\alpha_t}} \left(x_t^{ct} - \frac{\beta_t}{\sqrt{1 - \bar{\alpha}_t}} \epsilon_\theta(x_t^{ct}, x^{mask}, t) \right) + \sigma_t \mathbf{z}. \qquad (7)$$

Edge-Guided DDPM. Finally, rather than using only the semantic mask, we can leverage the semantic edge map, *e.g.*, x^{edge}, as the auxiliary information to guide the diffusion process. The network architecture was a combination of "conditional DDPM" and "Mask-guided DDPM", as shown in Fig. 2(c). Then in this case, Eq. (5) became:

$$x_{t-1}^{ct} = \frac{1}{\sqrt{\alpha_t}} \left(x_t^{ct} - \frac{\beta_t}{\sqrt{1 - \bar{\alpha}_t}} \epsilon_\theta(x_t^{ct} \oplus x^{mask}, x^{edge}, t) \right) + \sigma_t \mathbf{z}. \qquad (8)$$

Our implementation was based on [15]. Specifically, we set the time step $T = 1000$ and we used a linear noisy scheduler. The network used a ResNet backbone [6]. A hybrid loss function consisting of a L_2 loss item and variational lower-bound loss item was utilized to train the network. The optimizer was Adam with a learning rate of $1e^{-4}$. We trained the models for 150k iterations using a batch size of 16.

3   Experiments

Dataset. We used the training set of AMOS22 [9] CT subset to train all models. This training dataset consisted of 200 CT volumes of the abdomen from

200 different patients. The CT data was collected from multiple medical centers using different scanners, with detailed image acquisition information available in [9]. The dataset contained voxel-level annotations of 15 abdominal organs and structures: spleen, right kidney, left kidney, gallbladder, esophagus, liver, stomach, aorta, inferior vena cava, pancreas, right adrenal gland, left adrenal gland, duodenum, bladder, prostate/uterus. We added an additional "body" class to include the remaining structures beyond these 15 organs, and this was obtained through thresholding and morphological operations. The testing split comprised 50 CT volumes from 50 subjects that were taken from the AMOS22 CT validation set. We pre-processed the CT images by applying a windowing operation with a level of 40 and a width of 400. Each 2D slice in the 3D CT volume was extracted, normalized to the range of [0, 1], and resized to 256 × 256 pixels. This resulted in 26,069 images from 200 subjects for the training set, and 6559 images from 50 subjects for the testing set.

Baseline Comparisons. To comprehensively evaluate the performance of the proposed diffusion-based approaches, we compared them with several SOTA semantic image synthesis methods. These included GAN-based methods, such as SPADE [17], OASIS [19], Pix2Pix [8], as well as an existing diffusion-based approach SDM [22]. All comparative methods were implemented in PyTorch. The learning rates for SPADE and Pix2Pix were set at $5e^{-4}$. For OASIS, the learning rates were set at $4e^{-4}$ for the Discriminator and $1e^{-4}$ for the Generator. We used the Adam optimizer to train all models, for 300 epochs, with the most recent checkpoint used for evaluation. For the SDM model, we adhered to the same training scheme used for DDPM-based models.

Evaluation Metrics. We evaluated the performance based on both visual quality and learned organ correspondence. To assess visual quality, we used FID, SSIM, and PSNR. For assessing learned organ correspondence, we utilized an off-the-shelf, CT-only multi-organ segmentation network named TotalSegmentator (TS) [23]. This network was used to predict segmentation masks from synthetic CT volumes. Subsequently, Dice coefficients (DSC) were computed to compare these predicted masks with the ground-truth annotations.

4   Results and Discussion

Training Iteration Study. We initially evaluated the three different conditioning strategies after 50k, 100k, and 150k training iterations. Table 1 presents the numerical results. The overall trend indicated that the performance of all three proposed models converged after 150k training iterations. The mask-guided DDPM model outperformed the others by a small margin in most metrics at the 150k iteration mark. However, at earlier stages of training, specifically after 50k training iterations, the mask-guided DDPM model surpassed the conditional DDPM and edge-guided DDPM models, implying an earlier convergence. Furthermore, we observed that using auxiliary edge-map information did not improve performance. Figure 3 visualizes sample images after 10k, 50k, 100k, and 150k training iterations, respectively.

Table 1. FID, PSNR, SSIM, and DSC scores. The highest performance in each column is highlighted for different iterations setups.

Iter.	Methods	FID↓	PSNR↑	SSIM↑	DSC (%)↑ Sple.	Liv.	Kid_l	Kid_r	Panc.	Stom.	Aorta	Gall_bld	Espo.	Adr_r.	Adr_l.	Duod.	Cava.	Bladder
50k	Conditional DDPM	30.57	14.17	0.589	77.6	75.0	91.1	87.8	57.3	60.6	90.4	30.4	66.4	54.1	53.8	61.7	77.7	64.4
	Mask-guided DDPM	19.06	14.58	0.603	83.9	84.8	90.3	90.2	62.1	73.4	89.2	40.0	73.2	53.2	56.6	54.1	75.2	57.0
	Edge-guided DDPM	35.97	13.43	0.576	56.4	56.7	86.6	82.5	51.2	52.2	86.8	10.6	58.3	52.6	58.2	53.6	73.7	61.2
100k	Conditional DDPM	11.27	16.07	0.643	93.8	95.3	93.9	92.5	73.8	85.0	91.1	64.5	76.6	66.5	62.8	65.1	81.0	70.8
	Mask-guided DDPM	10.89	16.10	0.642	93.8	95.8	93.9	93.1	75.3	86.9	91.5	63.9	79.6	65.9	68.5	68.0	82.5	71.6
	Edge-guided DDPM	10.32	16.14	0.644	93.6	95.1	94.1	93.4	73.8	85.1	90.8	65.0	77.3	64.9	64.1	64.5	80.6	71.0
150k	Conditional DDPM	10.56	16.26	0.646	94.0	95.6	93.9	91.2	76.3	84.6	90.8	64.0	78.2	67.2	66.0	65.6	80.9	70.0
	Mask-guided DDPM	10.58	16.28	0.646	93.9	95.6	93.9	90.9	75.6	87.1	91.3	66.0	79.5	67.2	65.1	65.6	81.3	70.7
	Edge-guided DDPM	10.64	16.20	0.646	93.5	95.4	93.8	92.8	75.0	86.7	90.3	64.5	78.1	65.4	64.1	65.3	79.6	69.3

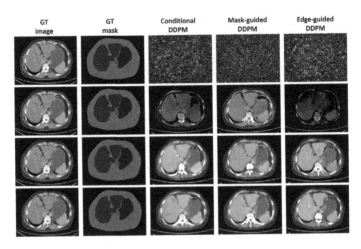

Fig. 3. Sampling results for three conditioning scenarios after 10k, 50k, 100k, and 150k training iterations. The first row shows the results after 10k training iterations; the second row shows results after 50k training iterations; the third row shows the results after 100k training iterations; the fourth row shows the results after 150k training iterations. The color map for different organs: liver (dark red), stomach (indigo), spleen (green), aorta (light brown), and inferior vena cava (aqua blue green). (Color figure online)

Comparison Study. We carried out a quantitative evaluation of the diffusion models against other SOTA algorithms such as Pix2Pix, OASIS, SPADE, and SDM methods, as shown in Table 2. In terms of image quality metrics such as FID, PSNR, and SSIM, the diffusion models outperformed non-diffusion-based methods. However, in terms of learned correspondence metrics like DSC, diffusion models surpassed other models for larger organs such as the liver, spleen, and kidneys. The OASIS method achieved superior performance for relatively small organs and structures like gallbladder and left adrenal gland. This may be because OASIS was good at synthesizing the clear boundary between small organs and the background, resulting in better segmentation results by TS and thus having higher DSC scores. Figure 4 presents multiple results ranging from the lower to the upper abdomen, from different methods. It is worth noting that from the top row of Fig. 4 GAN-based methods struggled to synthesize images when the number of mask classes was sparse. For example, Pix2Pix and SPADE were unable to generate realistic images. OASIS generated an image from the upper abdomen, which was inconsistent with the location of the given mask. The bottom row illustrated the same trend. GAN-based models failed to synthesize the context information within the body mask, for example, the heart and lung. On the contrary, diffusion models including the SDM model succeed to generate reasonable images based on the given masks. One explanation was that diffusion models were more effective when the number of masks decreased and the corresponding supervision became sparser.

Table 2. FID, PSNR, SSIM, and DSC scores for comparable methods. The highest performance in each column is highlighted.

Methods	FID↓	PSNR↑	SSIM↑	DSC(%)↑ Sple.	Liv.	Kid_l	Kid_r	Panc.	Stom.	Aorta	Gall_bld.	Espo.	Adr_r.	Adr_l.	Duod.	Cava.	Bladder
Pix2Pix	78.86	15.04	0.561	79.3	94.7	**94.5**	92.0	68.8	86.5	80.8	53.5	67.9	58.4	61.8	61.0	75.9	56.0
OASIS	43.57	14.75	0.560	91.8	93.5	92.0	88.1	73.4	**88.9**	88.6	**78.9**	**80.1**	**69.0**	**75.6**	**71.7**	**86.3**	**74.5**
SPADE	60.22	15.27	0.594	92.6	95.4	92.7	91.2	66.6	86.5	86.2	44.8	65.6	61.7	59.3	59.5	79.1	63.9
SDM	12.68	15.12	0.607	91.9	94.3	93.5	**93.2**	**79.3**	87.8	89.0	71.9	77.5	69.7	69.6	66.7	81.5	66.2
Conditional DDPM	**10.56**	16.26	**0.646**	**94.0**	**95.6**	93.9	91.2	76.3	86.4	90.8	64.0	78.2	67.2	66.0	65.6	80.9	70.0
Mask-guided DDPM	10.58	**16.28**	**0.646**	93.9	95.6	93.9	90.9	75.6	87.1	**91.3**	66.0	79.5	67.2	65.1	65.6	81.3	70.7
Edge-guided DDPM	10.64	16.20	**0.646**	93.9	95.4	93.8	92.8	75.0	86.7	90.3	64.5	78.1	65.4	64.1	65.3	79.6	69.3

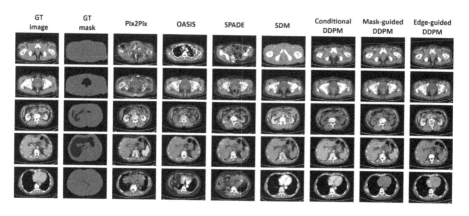

Fig. 4. Results from different semantic image synthesis methods. The color map for different organs: body (beige), spleen (green), liver (dark red), right kidney (blue), left kidney (yellow), stomach (indigo), aorta (light brown), duodenum (light purple), pancreas (gray), right/left adrenal gland (dark/light green), inferior vena cava (aqua blue green), bladder (shallow brown), prostate (purple). (Color figure online)

Future Work. One important application of generative models in medical imaging is to synthesize images for data augmentation. In the future work, we will use diffusion models as a data augmentation strategy and evaluate it in downstream segmentation, classification, or detection tasks. Compared with GAN-based generative models, the major limitation of diffusion models is that sampling procedures are more time-consuming and computationally expensive [10]. Nevertheless, multiple recent works successfully showed that using a reduced number of denoising steps was able to obtain high-quality samples, leading to faster inference procedures [21,26]. Therefore, by incorporating these techniques, we will investigate the role of conditional masks in fast sampling for synthesizing abdominal CT images.

5   Conclusion

In this work, we systematically investigated diffusion models for image synthesis for abdominal CT. Experimental results demonstrated that diffusion models outperformed GAN-based approaches in several setups. In addition, we also showed that disentangling mask and input contributed to performance improvement for diffusion models.

Acknowledgments. This work was supported by the Intramural Research Program of the National Institutes of Health (NIH) Clinical Center (project number 1Z01 CL040004). This work utilized the computational resources of the NIH HPC Biowulf cluster.

References

1. Chung, H., Ye, J.C.: Score-based diffusion models for accelerated MRI. Med. Image Anal. **80**, 102479 (2022)
2. Dhariwal, P., Nichol, A.: Diffusion models beat GANs on image synthesis. In: Advances in Neural Information Processing Systems, vol. 34, pp. 8780–8794 (2021)
3. Dorjsembe, Z., Pao, H.K., Odonchimed, S., Xiao, F.: Conditional diffusion models for semantic 3D medical image synthesis. arXiv preprint arXiv:2305.18453 (2023)
4. Fernández, R., Rosado, P., Vegas Lozano, E., Reverter Comes, F.: Medical image editing in the latent space of generative adversarial networks. Intell.-Based Med. **5** (2021)
5. Greenspan, H., Van Ginneken, B., Summers, R.M.: Guest editorial deep learning in medical imaging: overview and future promise of an exciting new technique. IEEE Trans. Med. Imaging **35**(5), 1153–1159 (2016)
6. He, K., Zhang, X., Ren, S., Sun, J.: Deep residual learning for image recognition. In: Proceedings of the IEEE Conference on Computer Vision and Pattern Recognition, pp. 770–778 (2016)
7. Hou, B.: High-fidelity diabetic retina fundus image synthesis from freestyle lesion maps. Biomed. Opt. Express **14**(2), 533–549 (2023)
8. Isola, P., Zhu, J.Y., Zhou, T., Efros, A.A.: Image-to-image translation with conditional adversarial networks. In: CVPR (2017)
9. Ji, Y., et al.: AMOS: a large-scale abdominal multi-organ benchmark for versatile medical image segmentation. arXiv preprint arXiv:2206.08023 (2022)

10. Kazerouni, A., et al.: Diffusion models in medical imaging: a comprehensive survey. Med. Image Anal., 102846 (2023)

11. Kim, B., Han, I., Ye, J.C.: DiffuseMorph: unsupervised deformable image registration using diffusion model. In: Avidan, S., Brostow, G., Cissé, M., Farinella, G.M., Hassner, T. (eds.) European Conference on Computer Vision, pp. 347–364. Springer, Cham (2022). https://doi.org/10.1007/978-3-031-19821-2_20

12. Lau, F., Hendriks, T., Lieman-Sifry, J., Sall, S., Golden, D.: ScarGAN: chained generative adversarial networks to simulate pathological tissue on cardiovascular MR scans. In: Stoyanov, D., et al. (eds.) DLMIA/ML-CDS -2018. LNCS, vol. 11045, pp. 343–350. Springer, Cham (2018). https://doi.org/10.1007/978-3-030-00889-5_39

13. Mahapatra, D., Bozorgtabar, B., Thiran, J.-P., Reyes, M.: Efficient active learning for image classification and segmentation using a sample selection and conditional generative adversarial network. In: Frangi, A.F., Schnabel, J.A., Davatzikos, C., Alberola-López, C., Fichtinger, G. (eds.) MICCAI 2018. LNCS, vol. 11071, pp. 580–588. Springer, Cham (2018). https://doi.org/10.1007/978-3-030-00934-2_65

14. Mok, T.C.W., Chung, A.C.S.: Learning data augmentation for brain tumor segmentation with coarse-to-fine generative adversarial networks. In: Crimi, A., Bakas, S., Kuijf, H., Keyvan, F., Reyes, M., van Walsum, T. (eds.) BrainLes 2018. LNCS, vol. 11383, pp. 70–80. Springer, Cham (2019). https://doi.org/10.1007/978-3-030-11723-8_7

15. Nichol, A.Q., Dhariwal, P.: Improved denoising diffusion probabilistic models. In: International Conference on Machine Learning, pp. 8162–8171. PMLR (2021)

16. Özbey, M., et al.: Unsupervised medical image translation with adversarial diffusion models. IEEE Trans. Med. Imaging (2023)

17. Park, T., Liu, M.Y., Wang, T.C., Zhu, J.Y.: Semantic image synthesis with spatially-adaptive normalization. In: Proceedings of the IEEE Conference on Computer Vision and Pattern Recognition (2019)

18. Saharia, C., Ho, J., Chan, W., Salimans, T., Fleet, D.J., Norouzi, M.: Image super-resolution via iterative refinement. IEEE Trans. Pattern Anal. Mach. Intell. (2022)

19. Schönfeld, E., Sushko, V., Zhang, D., Gall, J., Schiele, B., Khoreva, A.: You only need adversarial supervision for semantic image synthesis. In: International Conference on Learning Representations (2021). https://openreview.net/forum?id=yvQKLaqNE6M

20. Shin, H.-C., et al.: Medical image synthesis for data augmentation and anonymization using generative adversarial networks. In: Gooya, A., Goksel, O., Oguz, I., Burgos, N. (eds.) SASHIMI 2018. LNCS, vol. 11037, pp. 1–11. Springer, Cham (2018). https://doi.org/10.1007/978-3-030-00536-8_1

21. Song, J., Meng, C., Ermon, S.: Denoising diffusion implicit models. arXiv preprint arXiv:2010.02502 (2020)

22. Wang, W., et al.: Semantic image synthesis via diffusion models. arXiv preprint arXiv:2207.00050 (2022)

23. Wasserthal, J., et al.: TotalSegmentator: robust segmentation of 104 anatomic structures in CT images. Radiol. Artif. Intell., e230024 (2023). https://doi.org/10.1148/ryai.230024

24. Wolleb, J., Bieder, F., Sandkühler, R., Cattin, P.C.: Diffusion models for medical anomaly detection. In: Wang, L., Dou, Q., Fletcher, P.T., Speidel, S., Li, S. (eds.) International Conference on Medical Image Computing and Computer-Assisted Intervention, vol. 13438, pp. 35–45. Springer, Cham (2022). https://doi.org/10.1007/978-3-031-16452-1_4

25. Wolleb, J., Sandkühler, R., Bieder, F., Valmaggia, P., Cattin, P.C.: Diffusion models for implicit image segmentation ensembles. In: International Conference on Medical Imaging with Deep Learning, pp. 1336–1348. PMLR (2022)
26. Xiao, Z., Kreis, K., Vahdat, A.: Tackling the generative learning trilemma with denoising diffusion GANs. In: International Conference on Learning Representations (2022)
27. Zhao, X., Hou, B.: High-fidelity image synthesis from pulmonary nodule lesion maps using semantic diffusion model. In: Medical Imaging with Deep Learning, Short Paper Track (2023). https://openreview.net/forum?id=2M-2-75emE

CT Reconstruction from Few Planar X-Rays with Application Towards Low-Resource Radiotherapy

Yiran Sun[1], Tucker Netherton[2], Laurence Court[2], Ashok Veeraraghavan[1], and Guha Balakrishnan[1(✉)]

[1] Department of Electrical and Computer Engineering, Rice University, Houston, USA
{ys92,vashok,guha}@rice.edu
[2] Department of Radiation Physics, University of Texas MD Anderson Cancer Center, Houston, USA
{TNetherton,LECourt}@mdanderson.org

Abstract. CT scans are the standard-of-care for many clinical ailments, and are needed for treatments like external beam radiotherapy. Unfortunately, CT scanners are rare in low and mid-resource settings due to their costs. Planar X-ray radiography units, in comparison, are far more prevalent, but can only provide limited 2D observations of the 3D anatomy. In this work, we propose a method to generate CT volumes from few (<5) planar X-ray observations using a prior data distribution, and perform the first evaluation of such a reconstruction algorithm for a clinical application: radiotherapy planning. We propose a deep generative model, building on advances in neural implicit representations to synthesize volumetric CT scans from few input planar X-ray images at different angles. To focus the generation task on clinically-relevant features, our model can also leverage anatomical guidance during training (via segmentation masks). We generated 2-field opposed, palliative radiotherapy plans on thoracic CTs reconstructed by our method, and found that isocenter radiation dose on reconstructed scans have $<1\%$ error with respect to the dose calculated on clinically acquired CTs using ≤ 4 X-ray views. In addition, our method is better than recent sparse CT reconstruction baselines in terms of standard pixel and structure-level metrics (PSNR, SSIM, Dice score) on the public LIDC lung CT dataset. Code is available at: https://github.com/wanderinrain/Xray2CT.

Keywords: CT Reconstruction · Radiation Planning · Sparse Reconstruction · Deep Learning · Implicit Neural Representations

1 Introduction

CT scans are the standard-of-care for diagnosis and treatment of many diseases. However, due to their costs and infrastructure requirements, global inequities in access to CT scanners exist in many low-to-middle income countries (LMICs) [7]. This lack of CT access impacts many facets of healthcare such as external beam

A. Mukhopadhyay et al. (Eds.): DGM4MICCAI 2023, LNCS 14533, pp. 225–234, 2024.
https://doi.org/10.1007/978-3-031-53767-7_22

radiotherapy, in which a treatment planning system calculates the ionizing dose to a patient's tumor and surrounding tissues by utilizing the electron density information from the CT voxels. In comparison, planar X-ray units are far more prevalent in LMICs than CT units, and recent studies [16,22] demonstrate that significant information in CT scans may be estimated from sparse observations using deep generative networks trained over large datasets. With this motivation, we propose a learning-based algorithm for synthesizing CT volumes from few (<5) planar X-ray images, and demonstrate basic feasibility for radiotherapy planning for post-mastectomy chest walls (extremely prevalent for women in low-resource settings).

State-of-the-art CT reconstruction methods from sparse views are based on learning complex priors with neural networks and operate in both the sinogram [15,18], and voxel [6,9,16,22] spaces. Several voxel-based methods use convolutional neural networks (CNNs) optimized on (CT, X-ray) supervised pairs with <5 views [9,16,22]. Others use implicit neural representations (INRs), networks that map voxel coordinates to intensity values and can better reconstruct high-frequency details than CNNs [21]. However, INRs are typically fit using only the input views (i.e., self-supervised), and so require at least 20 views to attain reasonable results [24]. If such an approach were used with planar radiography, the large number of planar image acquisitions becomes practically infeasible, as technologists would need to reposition the patient and detector per orientation. In addition, previous studies provide limited evaluation, using only pixel-level reconstruction metrics like PSNR and SSIM [20].

To the best of our knowledge, we propose the first supervised CT reconstruction algorithm from few (<5) planar X-ray views using INRs. We build on the *pixelNeRF* [23] model design for sparse view synthesis problems. Our model first extracts 2D feature images from each input planar X-ray using a CNN U-Net [12]. For each 3D coordinate, it then uses an INR to predict the output CT's intensity given the coordinate, and a set of 2D features obtained by projecting the coordinate onto each feature image based on the known geometry of the X-ray imaging system. Our training loss function includes both a typical reconstruction term, and a segmentation term (captured by a pretrained segmentation network) which we hypothesize will be useful because radiotherapy plans rely on accurate anatomical boundaries.

We evaluated our method on reconstructing CT scans from 1 to 4 input planar X-rays. First, our method outperforms neural network baselines on the public LIDC-IDRI [1] lung CT dataset in terms of pixel-level (PSNR, SSIM) and structural (Dice Similarity Coefficient (DSC) [5]) metrics. Next, we evaluated our method using an in-house thoracic CT dataset for post-mastectomy chest wall radiotherapy, in which the tumor has been removed prior to the acquisition of the CT scan and the target of the radiotherapy would need only consider organs within the CT (e.g., chest wall, lungs, heart, spinal cord). 2-field opposed radiotherapy plans generated from our model's reconstructions obtain <1% error with respect to isocenter dose compared to clinical scans, well below the criterion for dose verification accuracy [25]. We conclude by discussing limitations and steps to move towards clinical application.

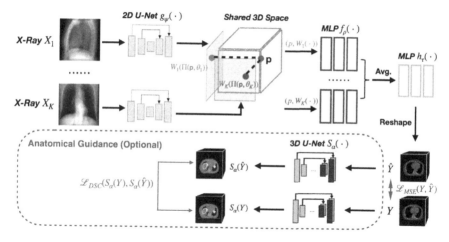

Fig. 1. Method overview. Our model takes different planar X-ray views as input and outputs a predicted CT volume $\hat{Y}$. 2D U-Net g_ψ generates a feature image W_i from each X_i. Then, for each 3D point $\mathbf{p}$, the projection operator Π retrieves the aligned feature vector $W_i(\Pi(\mathbf{p}, \theta_i))$ from each view and passes them into MLPs f_ρ and h_τ to predict intensity $\hat{Y}(\mathbf{p})$. We train the model with a reconstruction loss and an optional loss based on anatomical segmentation overlap (blue box). (Color figure online)

2   Method

Let $\mathbf{X} = \{X_1, \cdots, X_K\}$ represent K input planar X-rays acquired from different orientations $\{\theta_1, \cdots \theta_K\}$, where $X_i \in \mathbb{R}^{d \times d \times 1}$, and $Y \in \mathbb{R}^{d \times d \times d \times 1}$ represents the associated ground truth CT volume[1]. Our goal is to learn a model that maps $\mathbf{X}$ to Y. The main challenge of this reconstruction task is to combine the information from the different X-ray views into one shared 3D space. The overview of key components in our approach is illustrated in Fig. 1.

Building on *pixelNeRF* [23], we propose a model with three components: a 2D feature extraction network $g_\psi(\cdot)$ that extracts planar X-ray image features, a projection operation $\Pi(\cdot, \cdot)$ that maps 3D coordinates and a viewing angle to 2D coordinates, and an INR (implemented with functions $f_\rho(\cdot)$ and $h_\tau(\cdot)$) that maps 3D coordinates and K 2D feature vectors to voxel intensities. We supervise the entire model with a loss function consisting of a reconstruction error, and an (optional) segmentation error penalizing incorrect anatomical boundaries.

X-ray CNN: We implement function $g_\psi(\cdot)$ with a 2D CNN U-Net [12], which outputs an image $W_i \in \mathbb{R}^{d \times d \times c}$ encoding c multiscale features per pixel for X_i.

Projection: $\Pi(\cdot, \cdot) : \mathbb{R}^3 \times \mathbb{R}^1 \to \mathbb{R}^2$ maps a 3D coordinate and angle θ_i to the corresponding 2D location on image W_i based on the known X-ray imager geom-

---

[1] We assume a uniform dimension d here for simplicity, but our method can handle arbitrary dimensions.

etry. For example, if the X-rays were generated via parallel beam radiation, each point will be orthogonally projected onto W_i along angle θ_i. For fan-beam radiation, each point will be projected based on rays emanating from a 3D source point. The output of this operator is a feature vector $W_i(\Pi(\mathbf{p}, \theta_i)) \in \mathbb{R}^c$.

Conditional INR: Next, we use features $\{W_i(\Pi(\mathbf{p}, \theta_i))\}_{i=1}^K$ to estimate the voxel intensity at location $\mathbf{p}$. We use two multilayer perceptrons (MLPs) to do this: $f_\rho(\cdot, \cdot)$ and $h_\tau(\cdot)$. $f_\rho \in \mathbb{R}^2 \times \mathbb{R}^c \to \mathbb{R}^h$ operates on each view independently, and is responsible for combining a Fourier feature transform [19] of $\mathbf{p}$ and feature vector $W_i(\Pi(\mathbf{p}, \theta_i)))$ from view i into an embedding $\mathbf{r}_i(\mathbf{p})$. Fourier feature coordinate transforms empirically result in better high-frequency reconstructions compared to the coordinates on their own. Next, we compute the average embedding over all views $\hat{\mathbf{r}}(\mathbf{p})$, and feed it into MLP $h_\tau(\cdot)$, which outputs $\hat{Y}(\mathbf{p})$, an estimate of the intensity value (a scalar) at $\mathbf{p}$. We use three residual blocks for both MLPs, containing fully-connected linear layers with 128 neurons and sinusoidal periodic activation functions [17].

Loss Function: We train our model using the loss: $\mathcal{L}_{total} = \|\hat{Y} - Y\|_2^2 + \lambda \cdot \mathcal{L}_{DSC}(S_\alpha(\hat{Y}), S_\alpha(\hat{Y}))$, consisting of a typical mean squared error (MSE) term, and an (optional) term evaluating Dice score [5] between the segmentation masks of the two scans, estimated by pretrained segmentation network $S_\alpha(\cdot)$.

3   Experiments

We evaluated our model using the public Lung Image Database Consortium (LIDC-IDRI) [1] lung CT dataset, and an in-house Thoracic CT dataset from patients who received radiotherapy (gathered under an IRB approved protocol). LIDC includes 1018 patients, which we randomly split into 868/50/100 train/validation/test groups, and Thoracic includes 997 patients which we randomly split into 850/47/100 train/validation/test scans. We clipped all voxel values to $[-1000, 1000]$ Hounsfield Units (HU). We resampled each scan to 1 mm^3 resolution, cropped it to a cube, and then resized it to 128^3 voxels. We generated four planar X-ray views per CT at angles of: $0°$ (Lateral), $45°$, $90°$ (Frontal), and $135°$ using the Digitally Reconstructed Radiograph (DRR) generator *Plastimatch* [14], with energy level 50 keV. For our segmentation loss, we trained one segmentation network per dataset using a UNet [2,12]. For LIDC, we trained the segmentation network on 3 structures (left & right lung, nodule) using the LUNA16 [13] dataset. For Thoracic, we trained on 9 structures (see Fig. 3 for names) predicted by nnUNet network [8] used for contouring in the clinic. In the following results, we call our models trained without the segmentation loss *Ours*, and those trained with the segmentation loss *Ours-Seg*.

Metrics: We evaluated performance using three types of metrics: voxel level (PSNR, SSIM [20]), structural level (Dice Similarity Coefficient [5], or DSC), and radiation dose (Isocenter dose). Isocenter dose is defined as the calculated

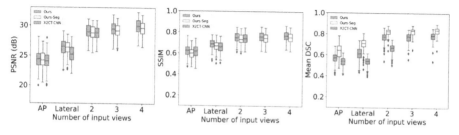

Fig. 2. Boxplots of PSNR, SSIM, and DSC between reconstructed and ground truth **CT** scans using **100 test patients in LIDC.** For 1 and 2 views, we also show performance of the baseline X2CT-CNN [22] (X2CT-CNN does not work with >2 views). Higher values are better.

dose (in centigray or cGy) deposited to a point in the patient's body at a distance of 100 cm away from a megavoltage X-ray source.

Baselines: We experimented with two neural network baselines: X2CT-CNN [22] and Neural Attenuation Fields (NAF) [24]. X2CT-CNN is a CNN for reconstructing CT scans from 1 or 2 (orthogonal) views. NAF is a recently proposed implicit neural representation (INR) that handles arbitrary viewing angles in CT reconstruction (like our model), but uses no prior training data. We trained both baselines from scratch on each dataset separately.

Implementation: We implemented our models in PyTorch [11] and ran all experiments on NVIDIA A100 GPUs with 40/80 GB of memory. We set the batch size to 1 and trained for 100 epochs per model. We used the ADAM [10] optimizer with an initial learning rate of $3e^{-5}$, and decreased the learning rate to $3e^{-6}$ after 50 epochs.

Radiotherapy Planning: Using 10 randomly selected patients from the Thoracic dataset, we generated radiotherapy plans with 2-field opposed beam arrangements using the RayStation commercial treatment planning system [3]. We set the isocenter within the thoracic spine of each clinical CT, fractional dose to 200 cGy, beam energy to 15 MV, and the radiation field size to $10 \times 10\,cm^2$ at isocenter. We performed rigid image registration so that radiation plans may be directly compared between the clinical and reconstructed CTs. Isocenter dose was compared between clinical plans and plans made on reconstructed CTs.

3.1   Results

First, we compare our models to baselines on LIDC using PSNR, SSIM, and DSC. NAF [24] performs poorly with a few number of views. For example, with 4 input views, 95% confidence intervals of PSNR, SSIM, and DSC are 23.23 ± 0.81, 0.440 ± 0.057, and 0.44 ± 0.05, respectively. Figure 2 shows the performance of our models and X2CT-CNN. Our models outperform X2CT-CNN for all views, with a particularly striking difference in DSC.

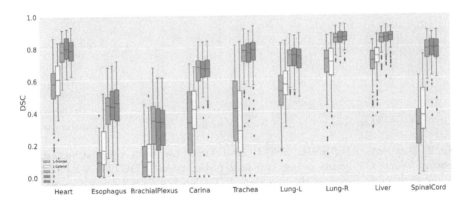

Fig. 3. Boxplot of Dice similarity coefficients (DSC) between reconstructed and ground truth CT scans using the Thoracic dataset. We compare versions of the proposed model Ours-Seg with different numbers of input views, on 100 test subjects. Higher values are better.

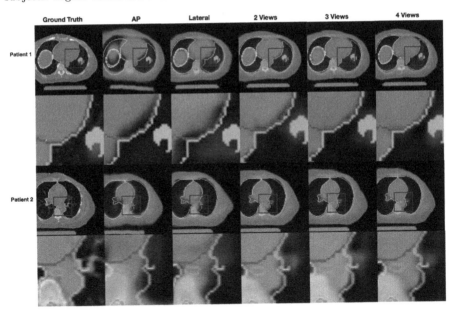

Fig. 4. Example reconstruction results on Thoracic dataset. We show results from two patients using the proposed model Ours-Seg. The first column shows the ground truth center slice. The remaining columns show the model's reconstructions for different numbers of input views. The pink contour segments the left lung, and the purple contour segments the heart.

Ground Truth   **2 Views**   **4 Views**

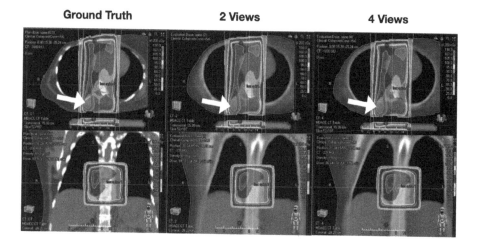

Fig. 5. Radiation planning visual results. Comparison of radiation plans between ground truth and reconstructed (columns 2–3) CTs for one patient using Ours-Seg. The top and bottom rows represent axial and coronal views of the given CT and overlaid dose distributions. Arrows in the top row indicate the (blue) region of maximum dose. Isodose lines closely match across all scans. (Color figure online)

Table 1. Average % errors of isocenter dose on 10 random subjects from Thoracic dataset. Our models obtain average errors under 1%. Standard deviations in parentheses.

	2 Views	4 Views
Ours	0.30 (0.35)	0.25 (0.26)
Ours-Seg	0.25 (0.26)	0.50 (0.57)

Moving from 1 to 2 views yields the largest marginal gains. *Ours-Seg* has higher DSC than *Ours*, but has slightly lower PSNR/SSIM. See Supplementary for a table with detailed results.

Next, Fig. 3 presents DSC boxplots of *Ours-Seg* with segmentation training on Thoracic. Again, the largest improvement occurs moving from 1 to 2 views, and the most difficult structures to contour are BrachialPlexus and Esophagus, likely because these structures are small in size. We also show sample reconstruction results with overlaid contours for two patients in Fig. 4 visually confirming the performance improvement near boundaries with more viewing angles.

Finally, Fig. 5 shows dose distribution results from treatment plans generated on reconstructed CTs using 2 and 4 input views. The shapes of the isodose lines closely resemble that of the ground truth, and in particular, for the high dose region (pointed to by white arrow). Additionally, average isocenter dose errors are under 1% (see Table 1), below the criterion for dose verification accuracy [25].

4   Discussion and Conclusion

Results demonstrate the feasibility of reconstructing CTs from few planar X-ray images. Segmentation guidance during training improves DSC (see Fig. 2), but did not have a consistent effect on isocenter dose error. The simple planning technique used in this work is used to treat regions in the spine and provide robustness against small uncertainties in patient position. Thus, the radiotherapy dose for this technique is not sensitive to small changes in CT voxel information, which may explain why segmentation-guided training had minimal effects. Further studies using complex, segmentation-driven treatment planning for multiple regions in the body would elucidate the relationship between subtle feature changes in the CT and its impact upon dose. Structural and dose level metrics presented here indicate that our approach also has potential for use with more complex treatments.

Results also show that our model combining a 2D CNN and INR is better for this task (in terms of voxel level metrics) than an INR only (NAF) that does not leverage prior training data, or a traditional CNN (X2CT-CNN) which suffers in modeling high-frequency details. Maximum performance gain occurs moving from 1 to 2 views, which makes sense since two orthogonal views are generally needed to confirm an object's location within the body [4].

Virtually all existing sparse CT reconstruction studies evaluate results using voxel-level metrics like PSNR and SSIM. This work makes the contribution of additionally evaluating in terms of radiotherapy plans. This is important, because all details need *not* be recovered for an algorithm to be clinically useful, a fact overlooked by PSNR and SSIM.

There are several exciting next steps to push this work forwards. First, CT reconstruction from few planar X-ray images is a highly ill-posed task and so there are infinitely many possible solutions per planar X-ray image input(s). By returning only one solution, our model is forced to produce scans that are the perceptual "average" of possible solutions. Incorporating a probabilistic formulation will help produce sharper results and quantify reconstruction uncertainties. Second, while the inference power of neural networks are remarkable, they are also known to "hallucinate" details. We will need further analysis into when and why such models make errors, with a particular focus on atypical subject cases. A focus of our next work will be 3-dimensional conformal radiotherapy planning for chest wall (post-mastectomy). For this specific use case, the tumor and diseased breast tissue is removed before the patient receives a CT and the target of the radiotherapy is the chest wall. Thus, tumor hallucination can be avoided and the development of such a technique will be extremely beneficial for women in low-resource settings, since post-masectomy radiotherapy is extremely prevalent. Finally, in practice we cannot assume that planar X-ray images are acquired at precise angles and depths. Development (and evaluation) of a model that can handle variable acquisition settings would therefore be a valuable contribution.

Acknowledgment. This work was supported by NSF CAREER: IIS-1652633.

References

1. Armato, S.G., III., et al.: The lung image database consortium (LIDC) and image database resource initiative (IDRI): a completed reference database of lung nodules on CT scans. Med. Phys. **38**(2), 915–931 (2011)
2. Balakrishnan, G., Zhao, A., Sabuncu, M.R., Guttag, J., Dalca, A.V.: VoxelMorph: a learning framework for deformable medical image registration. IEEE Trans. Med. Imaging **38**(8), 1788–1800 (2019)
3. Bodensteiner, D.: RayStation: external beam treatment planning system. Med. Dosim. **43**(2), 168–176 (2018)
4. Broder, J.: Imaging the chest: the chest radiograph. In: Broder, J. (ed.) Diagnostic Imaging for the Emergency Physician, pp. 185–296. W.B. Saunders, Saint Louis (2011). https://doi.org/10.1016/B978-1-4160-6113-7.10005-5. https://www.sciencedirect.com/science/article/pii/B9781416061137100055
5. Dice, L.R.: Measures of the amount of ecologic association between species. Ecology **26**(3), 297–302 (1945)
6. Ge, R., et al.: X-CTRSNet: 3D cervical vertebra CT reconstruction and segmentation directly from 2D X-ray images. Knowl.-Based Syst. **236**, 107680 (2022)
7. Hricak, H., et al.: Medical imaging and nuclear medicine: a lancet oncology commission. Lancet Oncol. **22**(4), e136–e172 (2021)
8. Isensee, F., Jaeger, P.F., Kohl, S.A., Petersen, J., Maier-Hein, K.H.: nnU-Net: a self-configuring method for deep learning-based biomedical image segmentation. Nat. Methods **18**(2), 203–211 (2021)
9. Jiang, Y., et al.: 3D volume reconstruction from single lateral X-ray image via cross-modal discrete embedding transition. In: Liu, M., Yan, P., Lian, C., Cao, X. (eds.) MLMI 2020. LNCS, vol. 12436, pp. 322–331. Springer, Cham (2020). https://doi.org/10.1007/978-3-030-59861-7_33
10. Kingma, D.P., Ba, J.: Adam: a method for stochastic optimization. arXiv preprint arXiv:1412.6980 (2014)
11. Paszke, A., et al.: PyTorch: an imperative style, high-performance deep learning library. In: Advances in Neural Information Processing Systems, vol. 32 (2019)
12. Ronneberger, O., Fischer, P., Brox, T.: U-Net: convolutional networks for biomedical image segmentation. In: Navab, N., Hornegger, J., Wells, W.M., Frangi, A.F. (eds.) MICCAI 2015. LNCS, vol. 9351, pp. 234–241. Springer, Cham (2015). https://doi.org/10.1007/978-3-319-24574-4_28
13. Setio, A.A.A., et al.: Validation, comparison, and combination of algorithms for automatic detection of pulmonary nodules in computed tomography images: the LUNA16 challenge. Med. Image Anal. **42**, 1–13 (2017)
14. Sharp, G.C., et al.: Plastimatch: an open source software suite for radiotherapy image processing. In: Proceedings of the XVI'th International Conference on the Use of Computers in Radiotherapy (ICCR), Amsterdam, Netherlands (2010)
15. Shen, L., Pauly, J., Xing, L.: NeRP: implicit neural representation learning with prior embedding for sparsely sampled image reconstruction. IEEE Trans. Neural Netw. Learn. Syst. (2022)
16. Shen, L., Zhao, W., Xing, L.: Patient-specific reconstruction of volumetric computed tomography images from a single projection view via deep learning. Nat. Biomed. Eng. **3**(11), 880–888 (2019)
17. Sitzmann, V., Martel, J., Bergman, A., Lindell, D., Wetzstein, G.: Implicit neural representations with periodic activation functions. In: Advances in Neural Information Processing Systems, vol. 33, pp. 7462–7473 (2020)

18. Sun, Y., Liu, J., Xie, M., Wohlberg, B., Kamilov, U.S.: CoIL: coordinate-based internal learning for tomographic imaging. IEEE Trans. Comput. Imaging **7**, 1400–1412 (2021)
19. Tancik, M., et al.: Fourier features let networks learn high frequency functions in low dimensional domains. In: Advances in Neural Information Processing Systems, vol. 33, pp. 7537–7547 (2020)
20. Wang, Z., Bovik, A.C., Sheikh, H.R., Simoncelli, E.P.: Image quality assessment: from error visibility to structural similarity. IEEE Trans. Image Process. **13**(4), 600–612 (2004)
21. Xie, Y., et al.: Neural fields in visual computing and beyond. Comput. Graphics Forum **41**, 641–676 (2022)
22. Ying, X., Guo, H., Ma, K., Wu, J., Weng, Z., Zheng, Y.: X2CT-GAN: reconstructing CT from biplanar X-rays with generative adversarial networks. In: Proceedings of the IEEE/CVF Conference on Computer Vision and Pattern Recognition, pp. 10619–10628 (2019)
23. Yu, A., Ye, V., Tancik, M., Kanazawa, A.: pixelNeRF: neural radiance fields from one or few images. In: Proceedings of the IEEE/CVF Conference on Computer Vision and Pattern Recognition, pp. 4578–4587 (2021)
24. Zha, R., Zhang, Y., Li, H.: NAF: neural attenuation fields for sparse-view CBCT reconstruction. In: Wang, L., Dou, Q., Fletcher, P.T., Speidel, S., Li, S. (eds.) Medical Image Computing and Computer Assisted Intervention-MICCAI 2022: 25th International Conference, Singapore, 18–22 September 2022, Proceedings, Part VI, vol. 13436, pp. 442–452. Springer, Cham (2022). https://doi.org/10.1007/978-3-031-16446-0_42
25. Zhu, T.C., et al.: Report of AAPM task group 219 on independent calculation-based dose/MU verification for IMRT. Med. Phys. **48**(10), e808–e829 (2021)

Reference-Free Isotropic 3D EM Reconstruction Using Diffusion Models

Kyungryun Lee and Won-Ki Jeong$^{(\boxtimes)}$

College of Informatics, Department of Computer Science and Engineering, Korea
University, Seoul, South Korea
{krlee0000,wkjeong}@korea.ac.kr

Abstract. Electron microscopy (EM) images exhibit anisotropic axial
resolution due to the characteristics inherent to the imaging modality, presenting challenges in analysis and downstream tasks. Recently
proposed deep-learning-based isotropic reconstruction methods have
addressed this issue; however, training the deep neural networks require
either isotropic ground truth volumes, prior knowledge of the degradation process, or point spread function (PSF). Moreover, these methods
struggle to generate realistic volumes when confronted with high scaling
factors (e.g. ×8, ×10). In this paper, we propose a diffusion-model-based
framework that overcomes the limitations of requiring reference data or
prior knowledge about the degradation process. Our approach utilizes
2D diffusion models to consistently reconstruct 3D volumes and is well-
suited for highly downsampled data. Extensive experiments conducted
on two public datasets demonstrate the robustness and superiority of
leveraging the generative prior compared to supervised learning methods.
Additionally, we demonstrate our method's feasibility for self-supervised
reconstruction, which can restore a single anisotropic volume without any
training data. The source code is available on GitHub: https://github.
com/hvcl/diffusion-em-recon.

Keywords: Diffusion models · Isotropic EM reconstruction ·
Super-Resolution

1 Introductions

While 3D electron microscopy (EM) provide exceptional lateral resolution of 3 to
5 nm per pixel, the prevalent technique of 3D EM imaging involves physically sec-
tioning tissue samples, resulting in a significantly lower axial resolution of approximately 30 to 50 nm per pixel (i.e., section thickness). This lower axial resolution
poses challenges particularly for small structures such as synaptic clefts that
can be smaller than the section thickness. [13] Conventional approaches, such

Supplementary Information The online version contains supplementary material
available at https://doi.org/10.1007/978-3-031-53767-7_23.

as interpolation and deconvolution, have been used to address this issue, offering fast solutions. However, these methods often produce unsatisfactory results, particularly when dealing with texture-rich EM images.

In recent years, deep learning-based methods have emerged as promising approaches for the isotropic reconstruction of EM images, outperforming the classical techniques. Heinrich et al. [8] leveraged isotropic FIB-SEM images to generate training data for supervised training of a 3D UNet-based super-resolution model. However, acquiring isotropic data in real-world scenarios is not feasible and there are cases where the downsampling process is unknown. On the other hand, by leveraging the point spread function (PSF), several studies [22,23] proposed a framework for fluorescence microscopy that does not require isotropic training data. By convolving the PSF with laterally viewed high-resolution image and subsampling, they simulated the anisotropic axial images. Training a 2D-UNet-like architecture using the generated pairs, they achieved superior performance compared to conventional deconvolution algorithms [14]. An interesting aspect of such approaches is the potential for self-supervised learning, as the target data itself can be used for training. Building upon this work, Deng et al. [5] conducted further experiments under more realistic settings. They demonstrated that self-supervised training using an inaccurate PSF could yield poor results. To address the limitation, they adopted a cycle-GAN [26] framework to implicitly learn the degradation process and generate proper low-resolution versions for training. Nevertheless, these works rely on deterministic reconstruction models that aim to minimize the pixel-wise error; hence, when performed on high-scaling factors ($\times 8$, $\times 10$), the results are blurry and fail to preserve fine structures.

In this study, to tackle a challenging scenario where no training data is available, we propose a novel approach that leverages the denoising diffusion probabilistic model (DDPM) [9] for realistic 3D EM reconstruction. The diffusion model is recently gaining attention due to its high fidelity and diverse generation compared to other generative models [6,17]. The diffusion model is adopted in various tasks for not only natural image domains [1,15,16] but also for medical image modalities [3,4,18,24]. They are also well known for handling inverse problems [2,10,20]. In particular, these methods are capable of restoring 2D images without requiring task-specific training or datasets. Consequently, considering that our 3D reconstruction problem can be regarded as a super-resolution (SR) task, which inherently is an inverse problem, we leverage diffusion models to address it. However, in the context of 3D generation, it becomes necessary to employ a generative model that can capture the underlying 3D data. Training such a model is challenging not only due to the significant memory resources required but also acquiring isotropic 3D training data is not feasible. Hence, we adopt a slice-by-slice approach by utilizing 2D diffusion models for the reconstruction of the 3D volume. We train a 2D diffusion model to learn the data distribution from high-resolution lateral images. Subsequently, we leverage the diffusion prior of the laterally trained model in the sequential reconstruction of the low-resolution axial images. In order to maintain coherence across the 3D

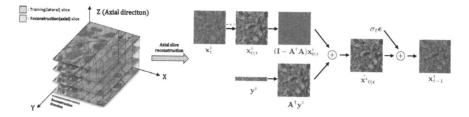

Fig. 1. (left): Reconstruction strategy for 3D EM via 2D diffusion models. The lateral images are used for training a diffusion model. Once trained, consistent sampling can be applied for any kind of 2D degradation (**A**). (right): Intuitive illustration of the refinement process. The low-frequency components are replaced with $\mathbf{A}^\dagger \mathbf{y}^i$ to fit the degradation process.

volume during the independent 2D reconstructions, we propose a sampling strategy where the previously reconstructed slice is encoded and used as a reference for the reconstruction of the next slice. We also introduce a heuristic method that improves the interpolated approximation to handle cases where the PSF is unknown, providing robustness in real-world scenarios. We validate the effectiveness and stability of the two proposed strategies via experiments and ablation studies. Our main contributions can be summarized as follows:

– We propose a sampling scheme that allows coherent 3D reconstruction using only 2D diffusion models. This allows smooth and continuous transitions between slices, therefore, eliminating artifacts when viewed in a perpendicular direction.
– We offer a simple but effective heuristic that can be applied without knowing the exact PSF which is often in practice. Moreover, the proposed approach is interpretable, allowing the reconstruction process to be more reliable.
– We demonstrate the superior reconstruction of DDPMs compared to previous auto-encoder-based methods by conducting simulation studies on a public dataset [19] for scenarios with/without prior information of the PSF. We also assess the performance on a real serial-section transmission electron microscopy (ssTEM) [7] volume without reference data or PSF information.

2   Method

As shown in Fig. 1(left), the whole process can be divided into two steps. We initially train a 2D DDPM on the lateral images of our target volume. This allows our generative model to learn "How high resolution images look like". Later, the laterally trained diffusion model is applied to reconstruct anisotropic axial planes slice-by-slice. Especially, we follow the restoring process of diffusion null-space model (DDNM) [20].

Preliminaries. Diffusion models first define a T-step forward process that progressively perturbates an image to pure noise $\mathbf{x}_T \sim \mathcal{N}(0, \mathbf{I})$ [9]. By defining the noise schedule parameters β_t, $\alpha_t := 1 - \beta_t$ and $\bar{\alpha}_t := \prod_{s=1}^{t} \alpha_s$, the forward process can be marginalized to a simple closed form of

$$q(\mathbf{x}_t|\mathbf{x}_0) = \mathcal{N}(\mathbf{x}_t; \sqrt{\bar{\alpha}_t}\mathbf{x}_0, (1 - \bar{\alpha}_t)\mathbf{I}), \tag{1}$$

The reverse process can be thought of as sampling from the posterior distribution, $q(\mathbf{x}_{t-1}|\mathbf{x}_t, \mathbf{x}_0)$. Therefore, to estimate the true posterior with $p_\theta(\mathbf{x}_{t-1}|\mathbf{x}_t)$, a noise predicting model is trained by minimizing the loss:

$$\mathcal{L}_{t-1} = \mathbb{E}_{\mathbf{x}_0,\epsilon,t}[||\epsilon - \epsilon_\theta(\sqrt{\bar{\alpha}_t}\mathbf{x}_0 + \sqrt{1 - \bar{\alpha}_t}\epsilon, t)||_2] \tag{2}$$

As described in DDIM [17], reparameterization allows inference in two steps by estimating the noise with ϵ_θ as follows:

$$\mathbf{x}_{0|t} = \frac{1}{\sqrt{\bar{\alpha}_t}}(\mathbf{x}_t - \epsilon_\theta(\mathbf{x}_t, t)\sqrt{1 - \bar{\alpha}_t}) \tag{3}$$

$$\mathbf{x}_{t-1} = \sqrt{\bar{\alpha}_{t-1}}\mathbf{x}_{0|t} + \sqrt{1 - \bar{\alpha}_t - \sigma_t^2} \cdot \epsilon_\theta(\mathbf{x}_t, t) + \sigma_t\epsilon, \quad \epsilon \sim \mathcal{N}(0, \mathbf{I}), \tag{4}$$

with $\sigma_t = \sqrt{(1 - \bar{\alpha}_{t-1})/(1 - \bar{\alpha}_t)}\sqrt{1 - \bar{\alpha}_t/\bar{\alpha}_{t-1}}$. Roughly speaking, at every iteration the reverse process is first estimating the clean image $\mathbf{x}_{0|t}$ at time t and again perturbing it with noise level $t - 1$, gradually decreasing the noise until $t = 0$.

DDNM [20] builds upon this inference process to solve linear inverse problems that are generally defined as $\mathbf{y} = \mathbf{A}\mathbf{x}$, where we aim to restore the data $\mathbf{x} \in \mathbb{R}^{n \times 1}$, given the degradation matrix $\mathbf{A} \in \mathbb{R}^{m \times n}$ and its observation $\mathbf{y} \in \mathbb{R}^{m \times 1}$. To sample an image that fits the constrain given by $\mathbf{A}$, range-space replacement is added after Eq. 3 as follows:

$$\hat{\mathbf{x}}_{0|t} = \mathbf{A}^\dagger\mathbf{y} + (\mathbf{I} - \mathbf{A}^\dagger\mathbf{A})\mathbf{x}_{0|t} \tag{5}$$

where $\mathbf{A}^\dagger \in \mathbb{R}^{m \times n}$ is the pseudo-inverse of $\mathbf{A}$ which can be calculated by the singular value decomposition (SVD) method. This refinement process ensures $\hat{\mathbf{x}}_{0|t}$ to satisfy the linear condition, hence shifting the direction of the reverse process to be consistent with the degradation. By substituting $\mathbf{x}_{0|t}$ with $\hat{\mathbf{x}}_{0|t}$ in Eq. 4, the noise mitigates the discrepancies between the replaced and original components of $\hat{\mathbf{x}}_{0|t}$. Figure 1(right) illustrates the refinement process for isotropic reconstruction.

Diffusion Models for 3D EM Reconstruction. As the degradation process occurs along the Z-axis of the 3D volume, it can be simplified as 2D degradations of contiguous ZY(or ZX) images. The 2D degradation can be represented with a matrix $\mathbf{A} = \mathbf{S}_f\mathbf{P}$, where $\mathbf{S}_f \in \mathbb{R}^{m \times n}$ is the sub-sampling operator choosing every f rows and $\mathbf{P} \in \mathbb{R}^{n \times n}$ is the PSF convolution operator. Assuming that we know the PSF and the downsampling factor, we can construct $\mathbf{A}$ and its

---

Algorithm 1. Reconstruction of the ith slice

Require: $\mathbf{y}^i$, $\mathbf{x}_0^{i-1}$, $\mathbf{A}$, $\mathbf{A}^\dagger$, ϵ_θ

  for $t = 0, ..., R - 1$ **do** $\triangleright$ Encode $\mathbf{x}_0^{i-1}$ deterministically

    $\mathbf{x}_{0|t}^{i-1} = \frac{1}{\sqrt{\bar{\alpha}_t}}(\mathbf{x}_t^{i-1} - \epsilon_\theta(\mathbf{x}_t^{i-1}, t)\sqrt{1 - \bar{\alpha}_t})$

    $\mathbf{x}_{t+1}^{i-1} = \sqrt{\bar{\alpha}_{t+1}}\mathbf{x}_{0|t}^{i-1} + \sqrt{1 - \bar{\alpha}_{t+1}} \cdot \epsilon_\theta(\mathbf{x}_t^{i-1}, t)$

  end for

  $\mathbf{x}_R^i = \mathbf{x}_R^{i-1}$

  for $t = R, ..., 1$ **do** $\triangleright$ DDNM reconstruction starting from $\mathbf{x}_R^{i-1}$

    $\epsilon \sim \mathcal{N}(0, \mathbf{I})$

    $\mathbf{x}_{0|t}^i = \frac{1}{\sqrt{\bar{\alpha}_t}}(\mathbf{x}_t^i - \epsilon_\theta(\mathbf{x}_t^i, t)\sqrt{1 - \bar{\alpha}_t})$

    $\hat{\mathbf{x}}_{0|t}^i = \mathbf{A}^\dagger \mathbf{y}^i + (\mathbf{I} - \mathbf{A}^\dagger \mathbf{A})\mathbf{x}_{0|t}^i$

    $\mathbf{x}_{t-1}^i = \sqrt{\bar{\alpha}_{t-1}}\hat{\mathbf{x}}_{0|t}^i + \sqrt{1 - \bar{\alpha}_t - \sigma_t^2} \cdot \epsilon_\theta(\mathbf{x}_t^i, t) + \sigma_t \epsilon$

  end for

---

pseudo-inverse $\mathbf{A}^\dagger$, therefore we directly apply the DDNM sampling procedure to the ith low-resolution ZY slice $\mathbf{y}^i$ to reconstruct $\mathbf{x}^i$. However, it is important to regard that the diffusion model used in the 2D reconstruction does not take into account the continuity between neighboring slices.

Therefore, we propose a consistent sampling strategy where the previous slice is encoded by DDIM and used as a starting point for the subsequent slice generation. This approach brings continuity between neighboring slices by leveraging the information encoded in the preceding slice. Moreover, most of the information overlaps between neighboring slices, therefore referencing the previous slice eases the reconstruction process for the diffusion model. Rather than beginning with pure Gaussian noise, we start the generation process of image $\mathbf{x}^i$ by encoding the previously reconstructed images $\mathbf{x}_0^{i-1}$ in a sequence of $[1, ..., R]$ and use $\mathbf{x}_R^{i-1}$ as a starting point. Specifically, by setting $\sigma_t = 0$ in Eq. 4 the DDIM iteration loses its stochasticity and it is possible to encode an image through a deterministic forward process. Given $\mathbf{x}_R^{i-1}$ and $\mathbf{y}^i$, the ith slice is reconstructed by the reverse process with $[R - 1, ..., 0]$, but this time with the introduction of random noise. As there is no previous reference for the first slice, the reverse diffusion process starts from Gaussian noise ($R = 1000$). The overall process is described in Algorithm 1. We also observed that, although our method allows smooth transition along the sampling axis, the perpendicular planes do not directly leverage the diffusion prior, thus showing unrealistic visual results. Therefore, we ensemble the two reconstructions processed along the x-axis and y-axis. Additionally, due to the generative model's stochastic nature, averaging the two results show a more steady and reliable generation.

In certain scenarios, the exact PSF is unknown. Therefore, we propose a simple approximation where we set $\mathbf{A}$ as linear down-sampling and $\mathbf{A}^\dagger$ as a linear interpolation operator. Despite the fact that linear interpolation is not the exact pseudo-inverse of linear down-sampling, [1] adopts it as a low-frequency guidance to generate an image in a desired direction. This approach enables the diffusion model to fill in the missing high-frequency details on top of the blurry interpolated observation $\mathbf{A}^\dagger \mathbf{y}$. As a result, the reconstructed data preserves the low-frequency structural information of the interpolated observation and remains interpretable without introducing abrupt changes. Although it has limitations that the reconstruction is a heuristic that relies on interpolation, we demonstrate through ablation that it gives better results compared to other assumptions of $\mathbf{A}$. We note that other kinds of interpolation methods can be used for $\mathbf{A}$ and $\mathbf{A}^\dagger$ instead of linear, e.g. cubic or lanczos.

3 Experiments

We assess the performance of our framework using two widely-used EM datasets: FIB-25 [19] and CREMI [7]. FIB-25 dataset is an isotropic FIB-SEM data commonly used for simulation studies, allowing quantitative evaluation of performance. CREMI is a ssTEM dataset with an anisotropic axial resolution. It serves as a real-world dataset for evaluating the performance of algorithms in handling anisotropic data. We trained the diffusion model with a U-Net backbone following [9] and adapted cosine scheduling [11] where $T = 1000$. The lateral training image size is 512×512 and the batch size is 4 with a learning rate of 0.00002. For sequential sampling, we reconstruct ZY images slice-by-slice along the x-axis, where the encoding/decoding level is $R = 200$. Except for the first axial slice, all slices were encoded by 4 steps and reconstructed (decoding) by 50 steps, where strided steps allow faster sampling. We compare our method with three auto-encoder-based approaches including 3D-SR-UNet [8], IsoNet [22,23] and the framework proposed by Deng et al. [5]. All the methods were implemented in Pytorch [12] and tested on a single NVIDIA RTX A6000 GPU.

Evaluation on Simulated Data. A randomly chosen subvolume of size $512 \times 512 \times 512$ from the isotropic FIB-25 data is convolved with a Gaussian filter and downsampled by choosing every $f \in \{4, 8\}$ lateral slices throughout the Z direction to generate synthetic anisotropic data. All deep-learning approaches except 3D-SR-UNet use only the target volume itself for training. 3D-SR-UNet is trained in a fully supervised manner with 26 additional isotropic subvolumes. Experiments are conducted for both cases where we know the PSF or not. For 3D-SR-UNet and IsoNet, we use average downsampling to generate low-resolution pair for blind-PSF. All methods require separate training for different PSF and downsampling factors, whereas our method does not require additional training. Performance is quantitatively evaluated by not only PSNR but also multi-scale structural similarity (MS-SSIM) [21] and LPIPS [25], which are more sensitive to fine patterns and structural details to demonstrate the realistic reconstruction achieved by our approach.

For $\sigma = 4$ and $f = 8$, Table 1 shows that our method outperforms all approaches for blind-PSF scenarios, especially in terms of LPIPS which measures the perceptual similarities between the reference and reconstructed images. In the case where we know the exact PSF, 3D-SR-UNet and IsoNet show better performance in PSRN and MS-SSIM. Nevertheless, the LPIPS score and

Table 1. Quantitative Comparison with other methods. With $f = 8$ and Gaussian filter of $\sigma = 4$, the isotropic FIB25 volume is simulated to an anisotropic resolution of $64 \times 512 \times 512$ The baseline is linear interpolation. $\perp$ and $+$ indicate that the reconstruction was done on ZX slices or is ensembled, respectively. The highest scores are highlighted in **bold**.

PSF	Method	ZY			ZX			XY		
		PSNR↑	MS-SSIM↑	LPIPS↓	PSNR↑	MS-SSIM↑	LPIPS↓	PSNR↑	MS-SSIM↑	LPIPS↓
	Baseline	26.12	0.842	0.567	26.11	0.840	0.555	26.18	0.848	0.379
Exact	3D-SR-UNet [8]	**28.96**	**0.934**	0.486	**28.97**	**0.931**	0.479	**29.04**	**0.931**	0.412
	IsoNet [22]	28.62	0.928	0.490	28.56	0.924	0.495	28.67	0.922	0.328
	Ours	27.92	0.914	**0.375**	27.93	0.916	0.434	28.03	0.913	0.296
	Ours⁺	28.39	0.924	0.426	28.39	0.922	**0.425**	28.49	0.920	**0.264**
Blind	3D-SR-UNet	27.52	0.894	0.512	27.52	0.891	0.503	27.57	0.895	0.426
	IsoNet	27.60	0.897	0.503	27.29	0.888	0.515	27.35	0.889	0.363
	Deng et al. [5]	27.65	0.901	0.496	27.65	0.901	0.504	27.75	0.900	0.408
	Ours	27.55	0.901	**0.391**	27.55	0.903	0.448	27.64	0.901	0.325
	Ours⊥	27.57	0.905	0.453	27.57	0.900	**0.393**	27.66	0.901	**0.280**
	Ours⁺	**27.95**	**0.911**	0.431	**27.95**	**0.909**	0.433	**28.04**	**0.908**	0.284

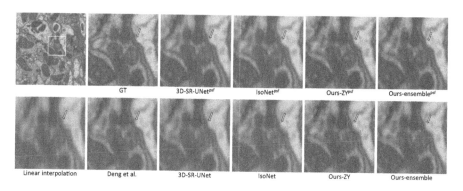

Fig. 2. Qualitative comparison of FIB25 reconstruction viewed in ZY. $f = 8$ and a Gaussian filter of $\sigma = 4$ was used. The superscript "psf" indicates that the exact point spread function was used for reconstruction.

visual results confirm that they cannot generate high-quality results. Moreover, 3D-SR-UNet is trained with isotropic volumes, thus incomparable. As discussed in Sect. 2, 'Ours⊥' shows that reconstruction along the ZX planes fails to generate realistic images viewed in ZY. Although the pixel-wise metrics preserve, the LPIPS score drops dramatically. 'Ours⁺' averages the two reconstructions along ZX and ZY, yielding a compromised result viewed in all directions and resulting in higher PSNR and MS-SSIM scores. Figure 2 shows that the autoencoder-based approaches tend to produce blurry results. This may be due to the limitation of the deterministic models based on pixel-wise loss functions, which may not fully capture the complexity and intricate details of the data. Furthermore, a noticeable quality gap exists between blind-PSF and exact-PSF cases for all methods except ours, which supports the robustness of our proposed heuristic. Experimental results for $\sigma = 2$ and $f = 4$ are described in Table S1 and Fig. S1.

Visual Comparison with Real ssTEM Data. In this section, we present reconstruction results of real anisotropic ssTEM data (CREMI) from $52 \times 512 \times 512$ to $512 \times 512 \times 512$, which does not have PSF information nor isotropic reference data. Thus, 3D-SR-UNet cannot be trained. For IsoNet, we use average downsampling to generate the low-resolution pairs for training. Our method uses the interpolation guidance method introduced in Sect. 2 for reconstruction. Figure 3 shows the reconstructed ssTEM volumes viewed in ZY and XY. IsoNet and Deng et al.'s method show blurry results for ZY. Moreover, the XY images show severe artifacts indicating misalignment.

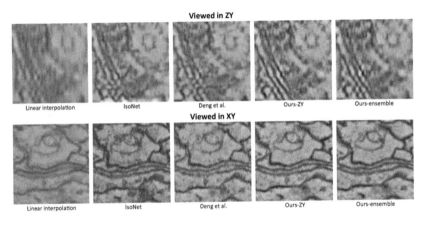

Fig. 3. Visual results of the reconstruction of a CREMI volume viewed in ZY and XY.

Ablation Studies. We perform two ablation studies on the FIB-25 dataset anisotropically simulated with $\sigma = 4$ and $f = 8$, starting with a comparison of different assumptions for the degradation process in blind-PSF scenarios. Imputation refers to the direct filling in of missing information, similar to the process of inpainting. We also compare it with average downsampling and an incorrect Gaussian filter of $\sigma = 2$ (note that $\sigma = 4$ is used to generate the synthetic anisotropic data). Table S2 and Fig. S2 suggests that reconstructing on top of the interpolated approximation gives better results than imputing or estimating the degradation using an incorrect filter. Secondly, we investigate the importance of continuous sampling throughout the reconstruction process. As shown in Fig. 4, when the previous slice is not encoded as a reference, the reconstruction exhibits visible artifacts in XY and ZX views.

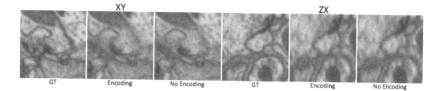

Fig. 4. Blind-PSF reconstructed volume viewed in XY and ZX. Reconstruction without referencing the previous slice shows severe artifacts due to the misalignment of adjacent slices.

4   Conclusion

We present a diffusion-model-based approach for reference-free isotropic reconstruction on highly anisotropic 3D EM volumes. We introduced two additional strategies that allow 2D diffusion models for consistent 3D reconstruction where the PSF is unknown. Through quantitative and qualitative results, we demonstrated its superiority compared to SOTA reconstruction methods and showed the limitations of auto-encoder-based frameworks. In addition to generating high-quality data, it exhibits efficacy in challenging conditions where training data is scarce and prior information is minimal, thereby demonstrating its potential in real-world applications. In the future, we plan to investigate the impact of our methods on various downstream tasks in biomedical domains.

Acknowledgements. This work was partially supported by the National Research Foundation of Korea (NRF-2019M3E5D2A01063819, NRF-2021R1A6A1A 13044830), the Institute for Information & Communications Technology Planning & Evaluation (IITP-2023-2020-0-01819), the Korea Health Industry Development Institute (HI18C0316), the Korea Institute of Science and Technology (KIST) Institutional Program (2E32210 and 2E32211), and a Korea University Grant.

References

1. Choi, J., Kim, S., Jeong, Y., Gwon, Y., Yoon, S.: ILVR: conditioning method for denoising diffusion probabilistic models. In: IEEE/CVF International Conference on Computer Vision (ICCV), pp. 14347–14356. IEEE (2021)
2. Chung, H., Kim, J., Mccann, M.T., Klasky, M.L., Ye, J.C.: Diffusion posterior sampling for general noisy inverse problems. arXiv preprint arXiv:2209.14687 (2022)
3. Chung, H., Ryu, D., McCann, M.T., Klasky, M.L., Ye, J.C.: Solving 3D inverse problems using pre-trained 2D diffusion models. In: Proceedings of the IEEE/CVF Conference on Computer Vision and Pattern Recognition, pp. 22542–22551 (2023)
4. Chung, H., Ye, J.C.: Score-based diffusion models for accelerated MRI. Med. Image Anal. **80**, 102479 (2022)
5. Deng, S., et al.: Isotropic reconstruction of 3D EM images with unsupervised degradation learning. In: Martel, A.L., et al. (eds.) MICCAI 2020. LNCS, vol. 12265, pp. 163–173. Springer, Cham (2020). https://doi.org/10.1007/978-3-030-59722-1_16
6. Dhariwal, P., Nichol, A.: Diffusion models beat GANs on image synthesis. In: Advances in Neural Information Processing Systems, vol. 34, pp. 8780–8794 (2021)

7. Funke, J., Saalfeld, S., Bock, D., Turaga, S., Perlman, E.: MICCAI challenge on circuit reconstruction from electron microscopy images. https://cremi.org/

8. Heinrich, L., Bogovic, J.A., Saalfeld, S.: Deep learning for isotropic super-resolution from non-isotropic 3D electron microscopy. In: Descoteaux, M., Maier-Hein, L., Franz, A., Jannin, P., Collins, D.L., Duchesne, S. (eds.) MICCAI 2017. LNCS, vol. 10434, pp. 135–143. Springer, Cham (2017). https://doi.org/10.1007/978-3-319-66185-8_16

9. Ho, J., Jain, A., Abbeel, P.: Denoising diffusion probabilistic models. In: Advances in Neural Information Processing Systems, vol. 33, pp. 6840–6851 (2020)

10. Kawar, B., Elad, M., Ermon, S., Song, J.: Denoising diffusion restoration models. In: Advances in Neural Information Processing Systems (2022)

11. Nichol, A.Q., Dhariwal, P.: Improved denoising diffusion probabilistic models. In: International Conference on Machine Learning, pp. 8162–8171. PMLR (2021)

12. Paszke, A., et al.: PyTorch: an imperative style, high-performance deep learning library. In: NeurIPS, pp. 8026–8037 (2019)

13. Plaza, S.M., Scheffer, L.K., Chklovskii, D.B.: Toward large-scale connectome reconstructions. Curr. Opin. Neurobiol. **25**, 201–210 (2014)

14. Richardson, W.H.: Bayesian-based iterative method of image restoration. JoSA **62**(1), 55–59 (1972)

15. Saharia, C., et al.: Palette: image-to-image diffusion models. In: ACM SIGGRAPH 2022 Conference Proceedings, pp. 1–10 (2022)

16. Saharia, C., Ho, J., Chan, W., Salimans, T., Fleet, D.J., Norouzi, M.: Image super-resolution via iterative refinement. IEEE Trans. Pattern Anal. Mach. Intell. (2022)

17. Song, J., Meng, C., Ermon, S.: Denoising diffusion implicit models. In: International Conference on Learning Representations (2020)

18. Song, Y., Shen, L., Xing, L., Ermon, S.: Solving inverse problems in medical imaging with score-based generative models. In: International Conference on Learning Representations (2022)

19. Takemura, S.Y., et al.: Synaptic circuits and their variations within different columns in the visual system of drosophila. Proc. Nat. Acad. Sci. **112**(44), 13711–13716 (2015)

20. Wang, Y., Yu, J., Zhang, J.: Zero-shot image restoration using denoising diffusion null-space model. arXiv preprint arXiv:2212.00490 (2022)

21. Wang, Z., Simoncelli, E.P., Bovik, A.C.: Multiscale structural similarity for image quality assessment. In: The Thrity-Seventh Asilomar Conference on Signals, Systems & Computers, vol. 2, pp. 1398–1402. IEEE (2003)

22. Weigert, M., Royer, L., Jug, F., Myers, G.: Isotropic reconstruction of 3D fluorescence microscopy images using convolutional neural networks. In: Descoteaux, M., Maier-Hein, L., Franz, A., Jannin, P., Collins, D.L., Duchesne, S. (eds.) MICCAI 2017. LNCS, vol. 10434, pp. 126–134. Springer, Cham (2017). https://doi.org/10.1007/978-3-319-66185-8_15

23. Weigert, M., et al.: Content-aware image restoration: pushing the limits of fluorescence microscopy. Nat. Methods **15**(12), 1090–1097 (2018)

24. Wolleb, J., Bieder, F., Sandkühler, R., Cattin, P.C.: Diffusion models for medical anomaly detection. In: Wang, L., Dou, Q., Fletcher, P.T., Speidel, S., Li, S. (eds.) Medical Image Computing and Computer Assisted Intervention-MICCAI 2022: 25th International Conference, Singapore, 18–22 September 2022, Proceedings, Part VIII, vol. 13438, pp. 35–45. Springer, Cham (2022). https://doi.org/10.1007/978-3-031-16452-1_4

25. Zhang, R., Isola, P., Efros, A.A., Shechtman, E., Wang, O.: The unreasonable effectiveness of deep features as a perceptual metric. In: Proceedings of the IEEE Conference on Computer Vision and Pattern Recognition, pp. 586–595 (2018)
26. Zhu, J.Y., Park, T., Isola, P., Efros, A.A.: Unpaired image-to-image translation using cycle-consistent adversarial networks. In: IEEE International Conference on Computer Vision (ICCV), pp. 2242–2251. IEEE (2017)

Author Index

A. Mukhopadhyay et al. (Eds.): MICCAI 2023, LNCS 14533, pp. 247–248, 2024.
https://doi.org/10.1007/978-3-031-53767-7

Printed in the United States
by Baker & Taylor Publisher Services